DRONE

PHOTOGRAPHY & VIDEO MASTERCLASS

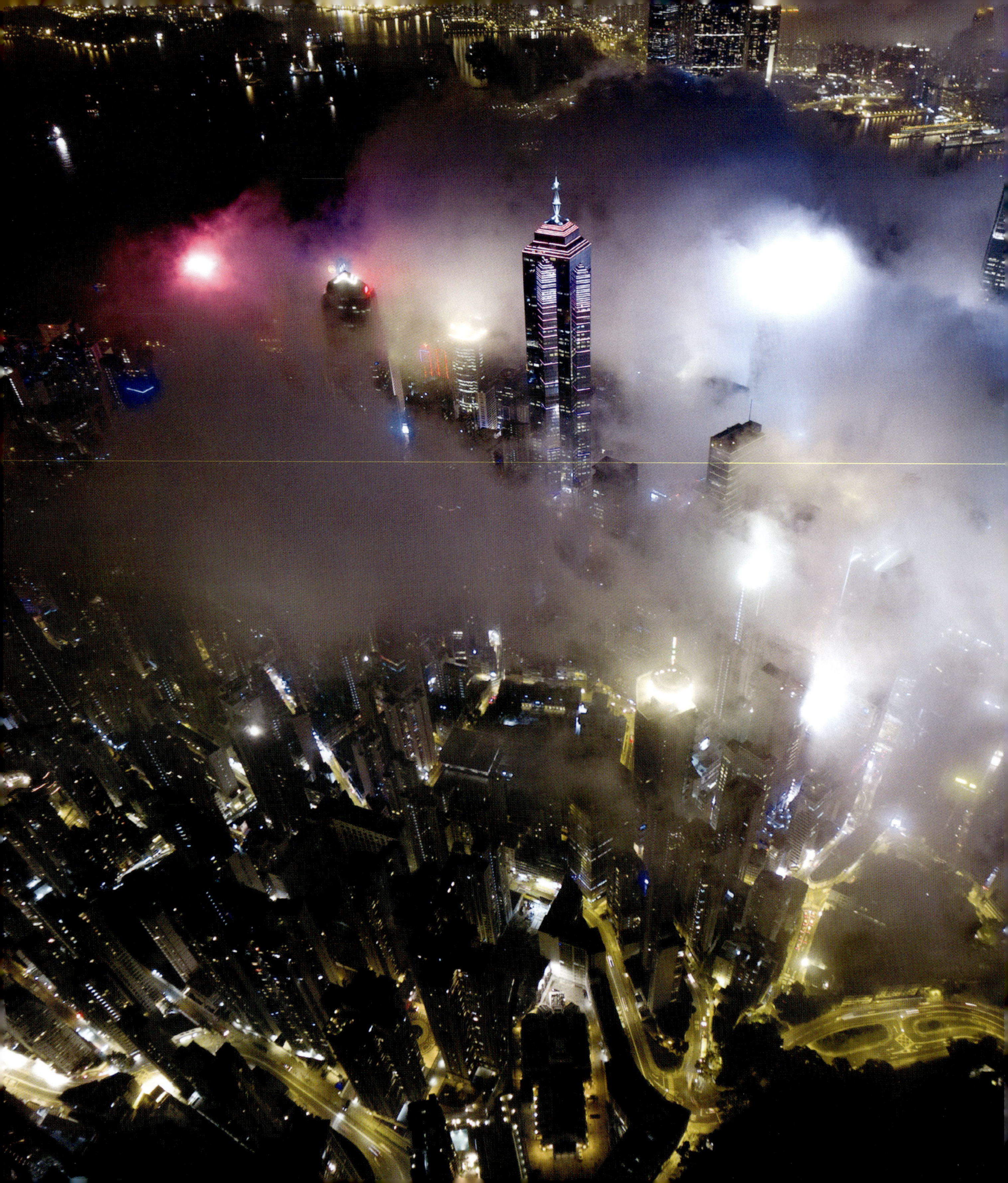

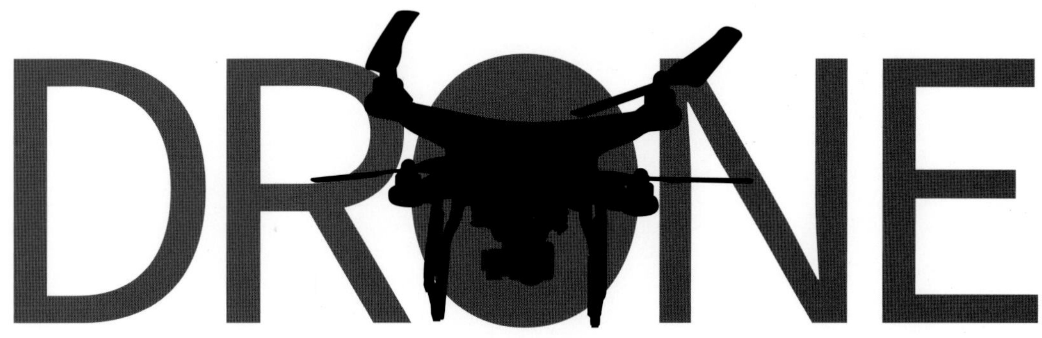

PHOTOGRAPHY & VIDEO MASTERCLASS

FERGUS KENNEDY

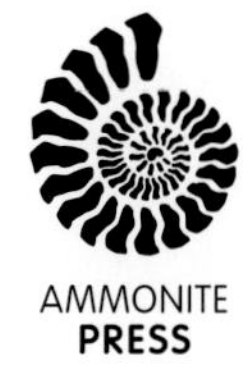

First published 2017 by Ammonite Press
an imprint of Guild of Master Craftsman Publications Ltd
Castle Place, 166 High Street, Lewes, East Sussex, BN7 1XU,
United Kingdom

ISBN 978 1 78145 300 1

A catalog record for this book is available from the
British Library.

Publisher: Jason Hook
Art Director: Robin Shields
Editor: Chris Gatcum
Designer: Luke Herriott

Color reproduction by GMC Reprographics
Printed and bound in Malaysia

CONTENTS

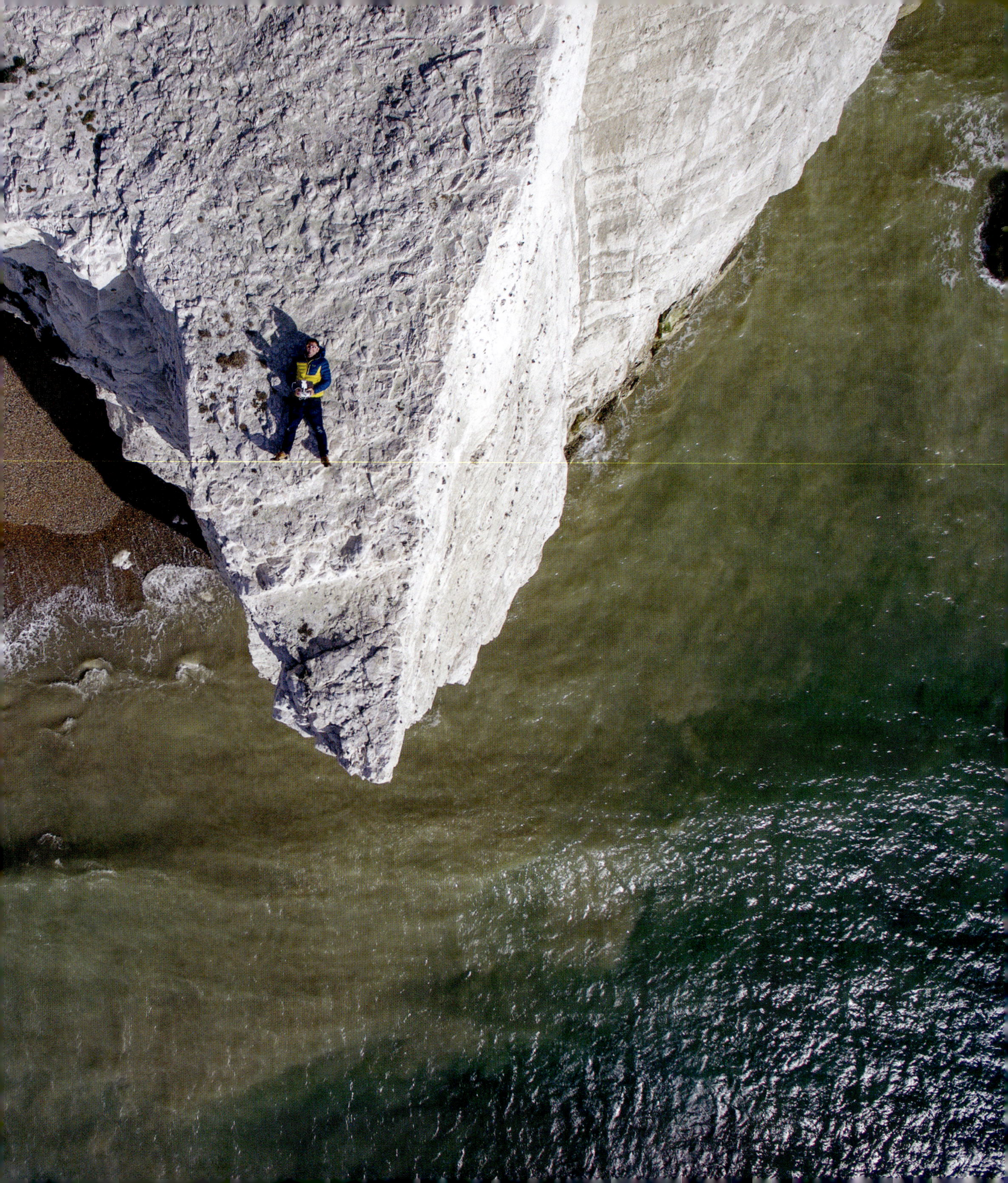

INTRODUCTION

If all of your photography takes place with your feet planted firmly on the ground, you're missing out on a whole dimension! The ability to look down on the earth, as though through the eyes of a bird, has been a dream of mankind throughout history. I've often wished I could get an aerial view of a particular scene—having worked as a photographer from a variety of full-sized aircraft, I knew what could be achieved from the air, but it was often too expensive or impractical. So when affordable camera drones started to become available, I was really excited to become involved—the air above and around us now represents a realm of photographic opportunity that is rapidly becoming more accessible.

But how best to approach this amazing new opportunity? The ideas and tips put forward in this book can be summarized by the phrase "You're not flying a drone, you're moving a camera." By this, I don't mean that you can ignore all the skills and safety aspects involved in operating an unmanned aerial vehicle (UAV), it is just that these should become second nature. Once that happens, you can concentrate on the photographic and creative elements involved in making beautiful images.

Whether you are a keen photographer in search of new frontiers, or you have a drone and want to hone your photographic or video skills, this book contains all the guidance and inspiration you need. The amount of information may seem daunting at first, but if you fly your drone with specific goals in mind—concentrating on one technique at a time and carefully planning your flight toward a specific shot—you will find that your photography or video skills improve at a much faster rate. You are far better off capturing one stunning shot on a flight than snapping dozens of mediocre ones!

Although there is plenty of detail in this book about equipment—and you certainly need to know your way around your kit—it's important to realize that it is the person operating the controls that makes the photograph. It's very easy to become distracted by the equipment, but the magic happens when the drone becomes an extension of you and you can concentrate on the view through the camera lens. Most of the outstanding drone shots I have seen have been taken by talented and experienced photographers using very modest equipment, and you can quite easily join their ranks.

◀ Seeing the world through the eyes of a bird. Flying a drone for photography is a fascinating and absorbing pastime.

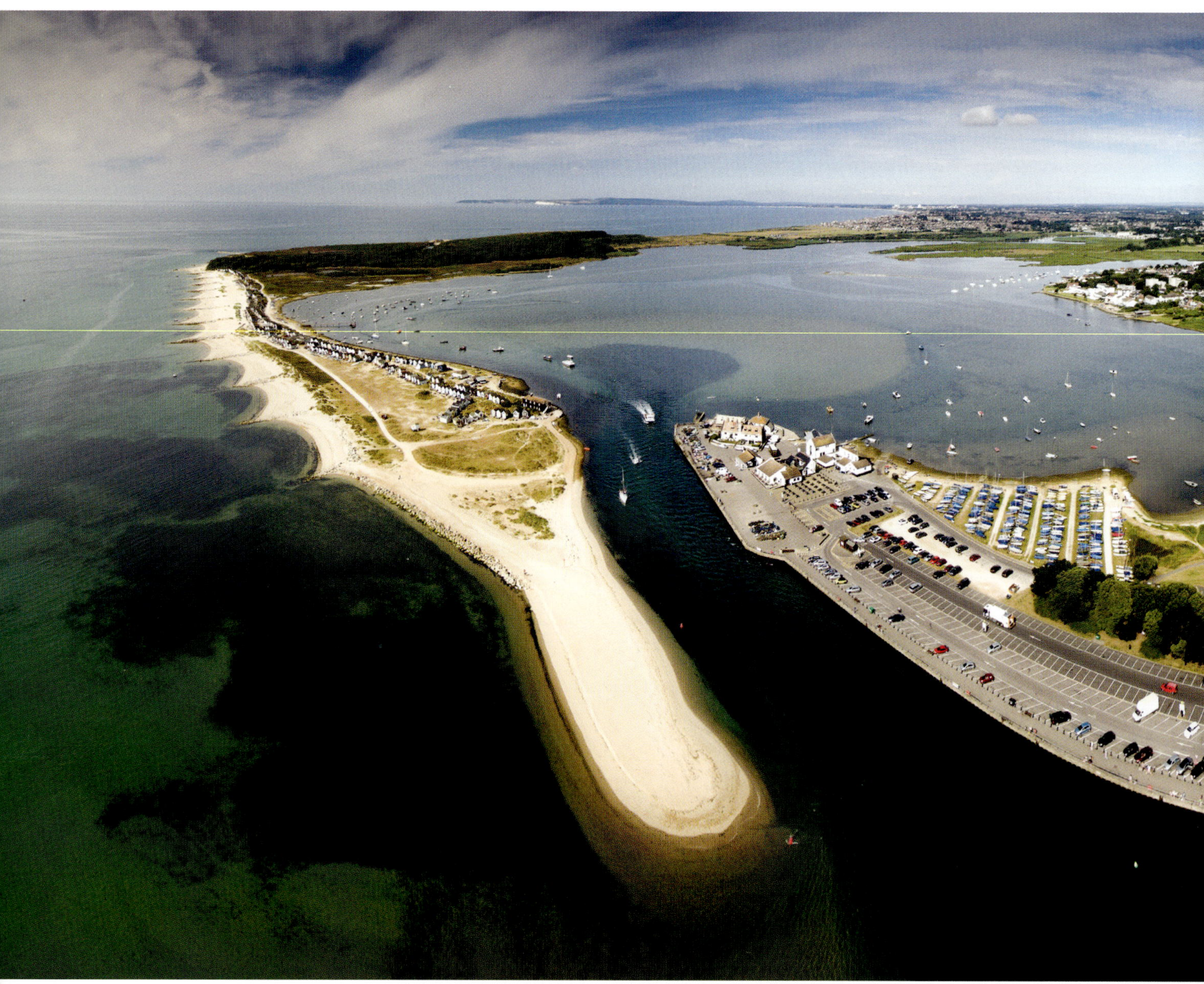

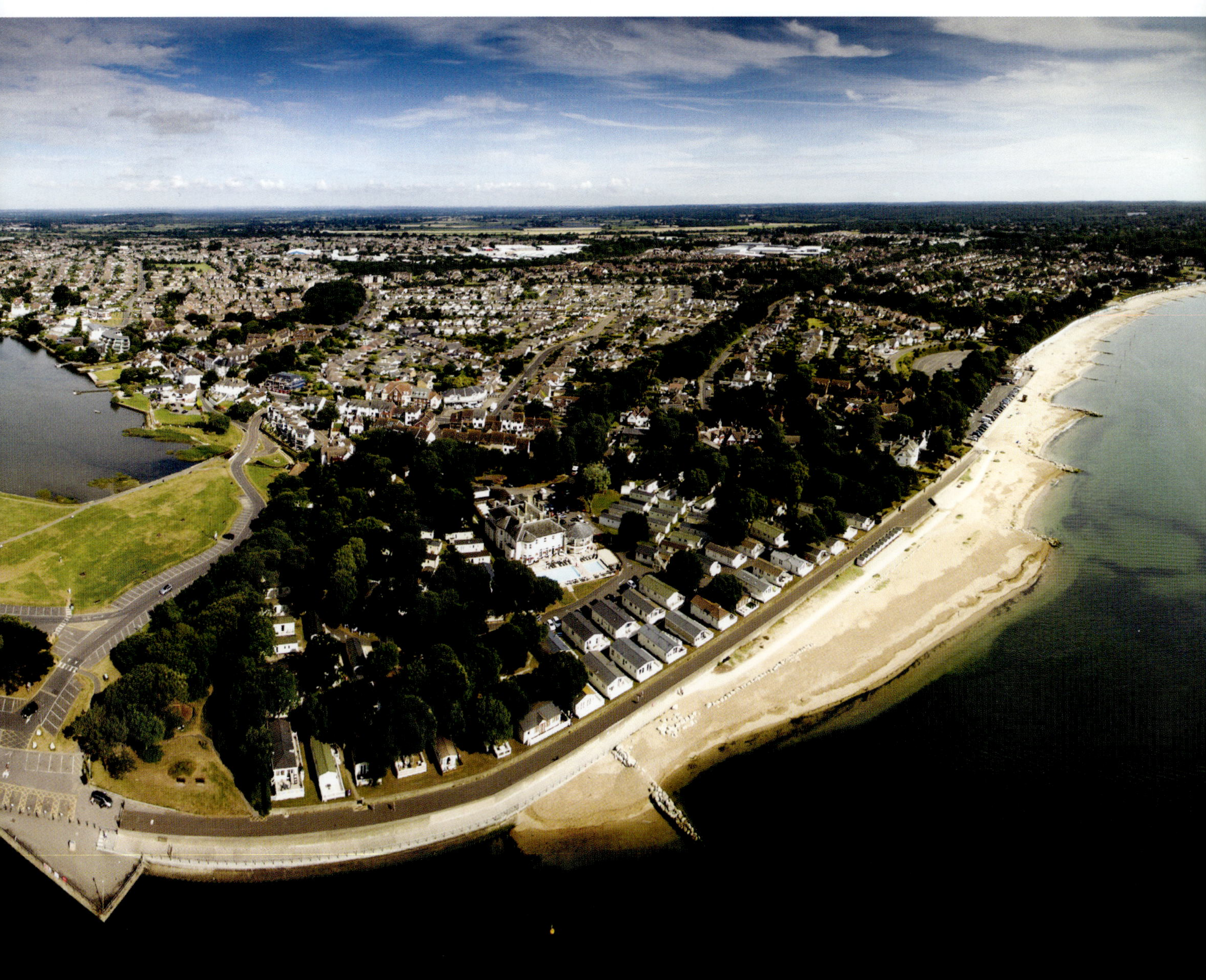

▲ Christchurch Harbor, Dorset, UK. This stitched panoramic shot was created from multiple photographs taken using a DJI Phantom 3 Pro.

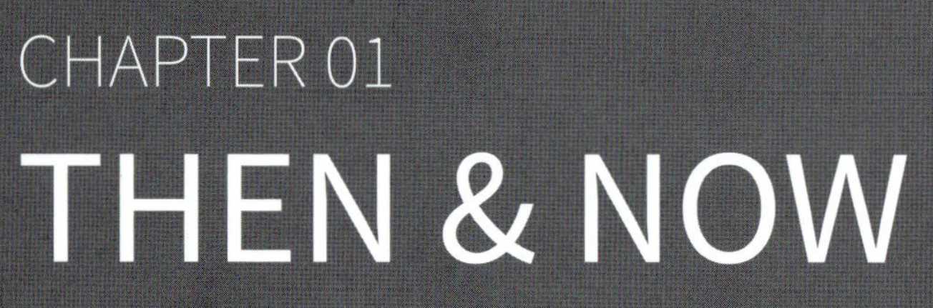

CHAPTER 01

THEN & NOW

Aviation has come a long way in the past 100 years, with aerial photography taking similarly huge strides alongside it. Many ingenious methods have been employed to lift cameras into the skies, both with and without a human operator. Modern drone photography has made this easier than ever, so today's photographers can reap the rewards of technological leaps in aviation, flight control electronics, and cameras.

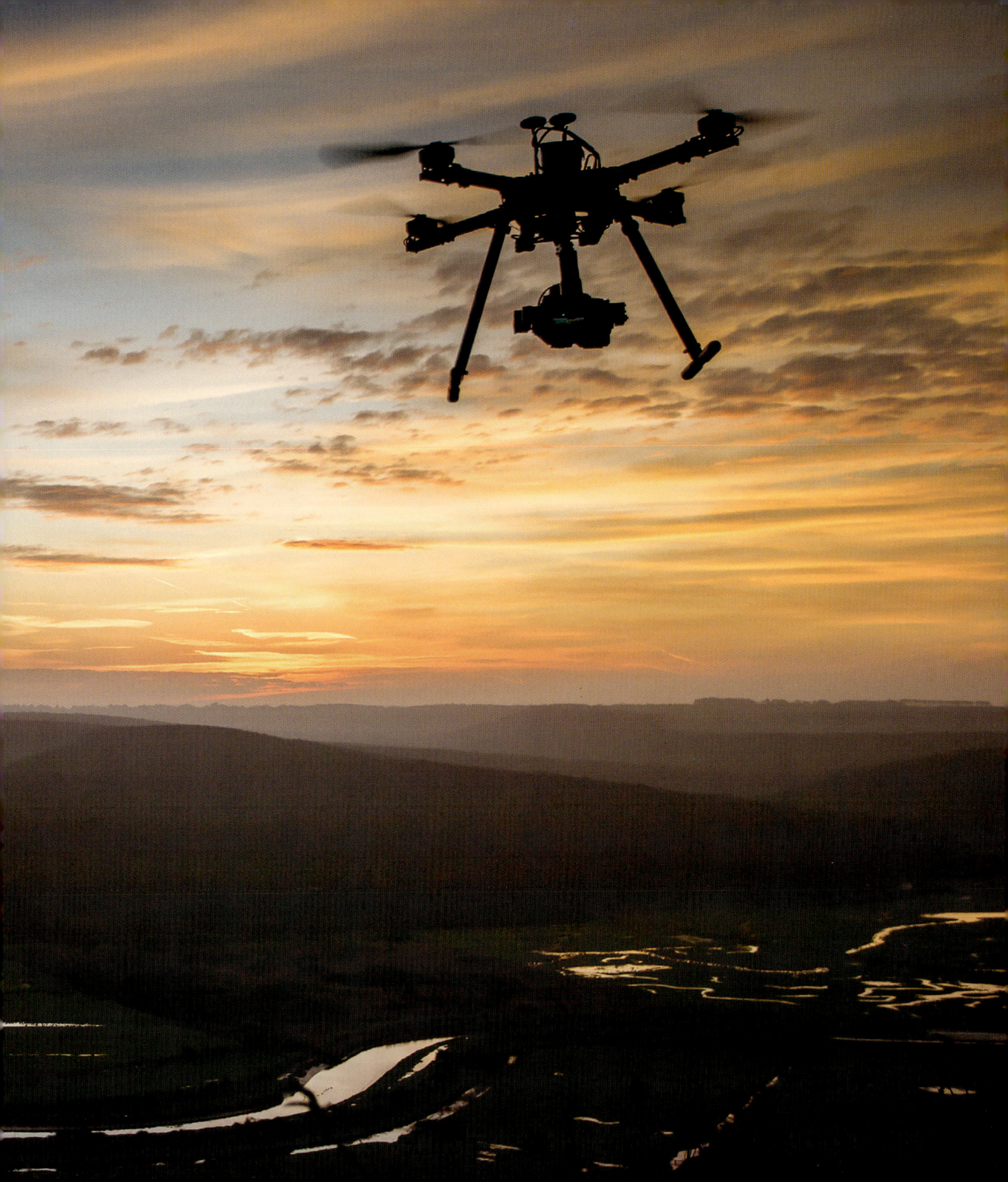

EARLY AERIAL PHOTOGRAPHY

Historically, the reasons for wanting to get a camera aloft were many and varied. Although some were purely artistic, many early attempts had practical intentions, such as military intelligence, mapping, or surveying.

Pioneering aerial photographers had already found ways to get their cameras into the air long before the Wright Brothers made the first powered flight in 1903: some attempts were more successful than others and they ranged from the ambitious to the downright bizarre.

▼ The first aerial photographs were taken from hot air balloons in the second half of the 19th century, led by the famous French photographer, Nadar.

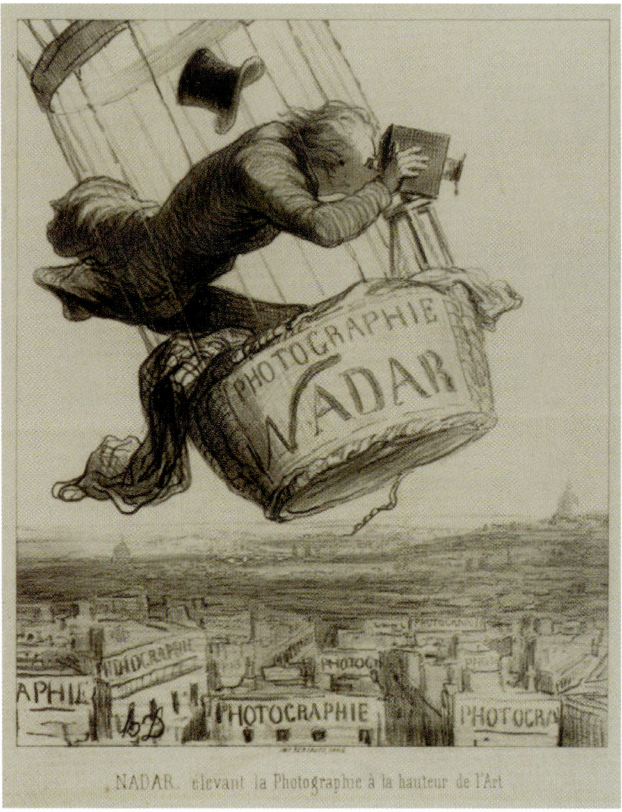

The earliest aerial photography arguably took place in 1858, when the French photographer, Gaspard-Félix Tournachon (better known as Nadar), took photographs from a hot air balloon. Balloons are a fantastically stable, almost magical platform for aerial photography, but they have numerous problems: they are bulky, the pilot has limited control over their direction of flight, and they can only be used on very calm days.

In 1882, the British meteorologist, E. D. Archibald, was among the first to start using a kite to lift a camera. Since then, kites have been deployed successfully for photographic purposes, including photographing the aftermath of the devastating San Francisco earthquake of 1906. Kite aerial photography continues to have a small (but enthusiastic) band of devotees.

Around the same time as Archibald was attaching cameras to kites, a German apothecary named Julius Gustav Neubronner was experimenting with using pigeons as aerial photographers. This saw small cameras fitted to breastplates on pigeons, with photographs taken using a timer delay. Although Neubronner made some successful aerial shots with his pigeons, the main problem was a lack of control over the subject matter. It must also have been slightly nerve-racking waiting for the pigeon and its expensive camera gear to return home!

In 1906, another German, Albert Maul, used a compressed-air-powered rocket to shoot a camera 2600ft (almost 800m) into the air. This approach was refined through the use of powder-propelled rockets and gyroscopically stabilized cameras, but by the time it was successfully demonstrated to the Austrian Army in 1912, a new way of lifting cameras was already taking over.

Aerial photography really hit the mainstream when manned, fixed-wing aircraft became a viable option. Just six years after his maiden flight, Wilbur Wright took the first photographs from a manned, fixed-wing aircraft in 1909, with aerial photographs forming the basis of battlefield maps during World War I.

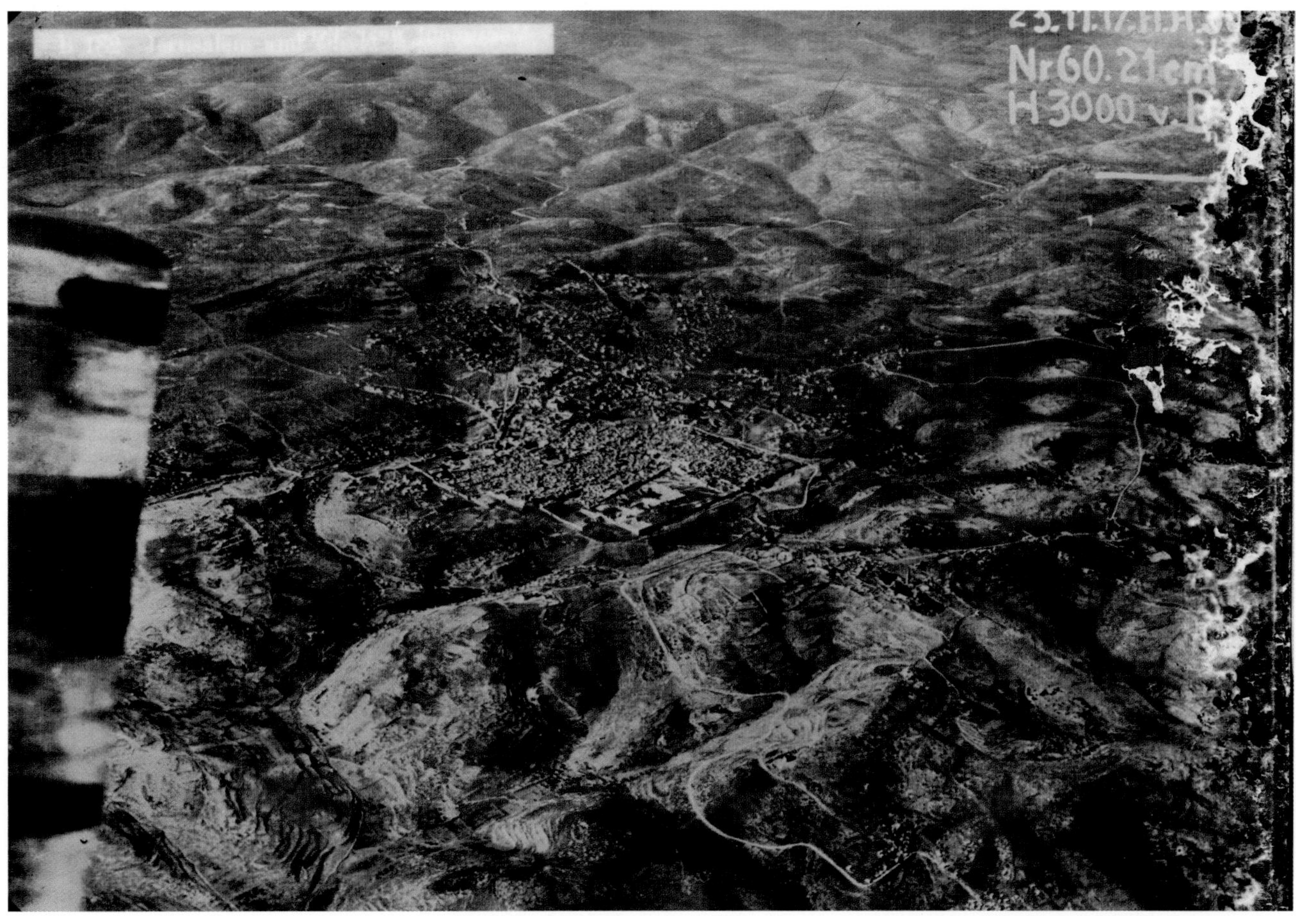

▲ As fixed-wing aircraft became more widely used, and dry plate photography became more widespread, cameras could more easily take to the skies. This photograph of Jerusalem and Olivet, dating from around 1900, was among the first aerial photographs taken of Palestine and Syria. It was taken from an altitude of almost 10,000ft (3000m) and recorded on a 5x7in glass plate.

However, various tweaks needed to be made to optimize the cameras for aerial photography. One of the pioneers in this field was Sherman Fairchild, who graduated from Harvard in 1915. Within a couple of years he was designing cameras to be fitted to early biplanes. His designs saw the camera's shutter integrated into the lens, which greatly reduced camera shake problems. This would become the standard in aerial photography for the next half century.

Fairchild later branched out into aircraft design after realizing that existing designs were not ideal for photography. Perhaps his most remarkable achievement was when he saw his camera designs go into space on the Apollo missions to photograph the moon's surface from lunar orbit. This set the scene for the increasingly sophisticated photographic equipment found in high-altitude spy planes and satellites.

DAWN OF THE DRONES

The development of the modern camera drone has put a highly effective aerial photography tool into the hands of a great many more enthusiasts, without the need for astronomical amounts of money and technical knowledge. Although enthusiast and professional photographers have been successfully fitting cameras to a range of radio-controlled airplanes and helicopters for many years, it is only recently, as a result of advances in multirotor technology that radio-controlled aerial photography has become truly accessible.

There is an ongoing debate about the definition of the term "drone." Originally, it was used to refer to any unmanned aircraft that could operate autonomously or beyond the sight of the operator, but the word has been "tainted" slightly by close links to weaponized military drones. Negative connotations also seem to stem from the perception that Peeping Toms and prying government agencies use drones. Consequently, some people prefer to use the term "unmanned aerial vehicle" (or UAV) to distinguish their "civilian" remotely operated multirotor aircraft with cameras from their slightly more insidious brethren. However, there is no denying that these aircraft are commonly known as drones, which is the term we shall use here.

Broadly speaking, the aircraft we are interested in can be divided into two types: multirotor and fixed wing. The setup you choose will depend on several factors, including your budget, the technical requirements of the camera, your experience level, and portability. To find rare animals in a remote area you will need a good range on the aircraft, a long flight time, and a powerful zoom on the camera. However, high operating costs will limit its use to the biggest budget productions.

At the other end of the spectrum, a small and relatively cheap drone can fit into a small backpack and go with you almost anywhere. This type of aircraft is quick to deploy and it's now feasible to take a small drone with you alongside conventional photography equipment, even on an extended trek.

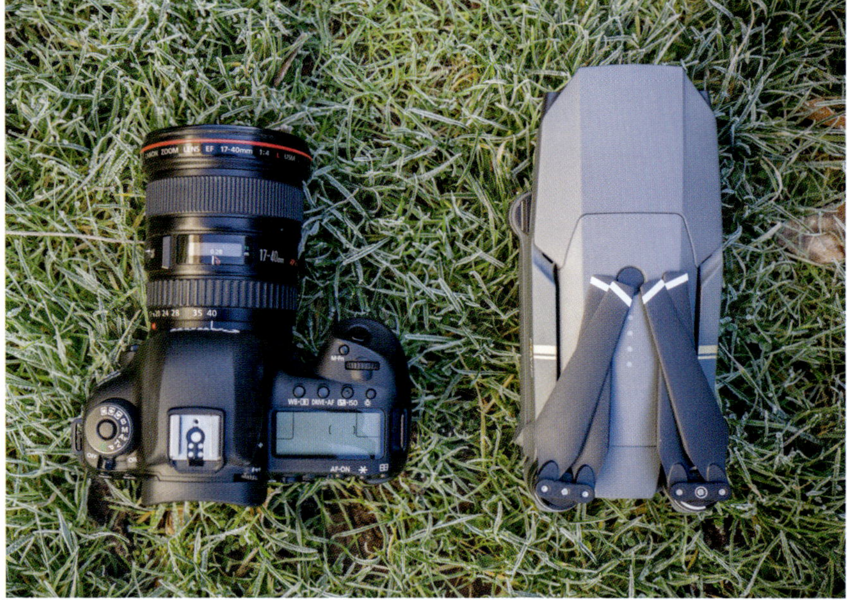

◀ A modern folding camera drone can take up roughly the same amount of space in your bag as an SLR camera and lens.

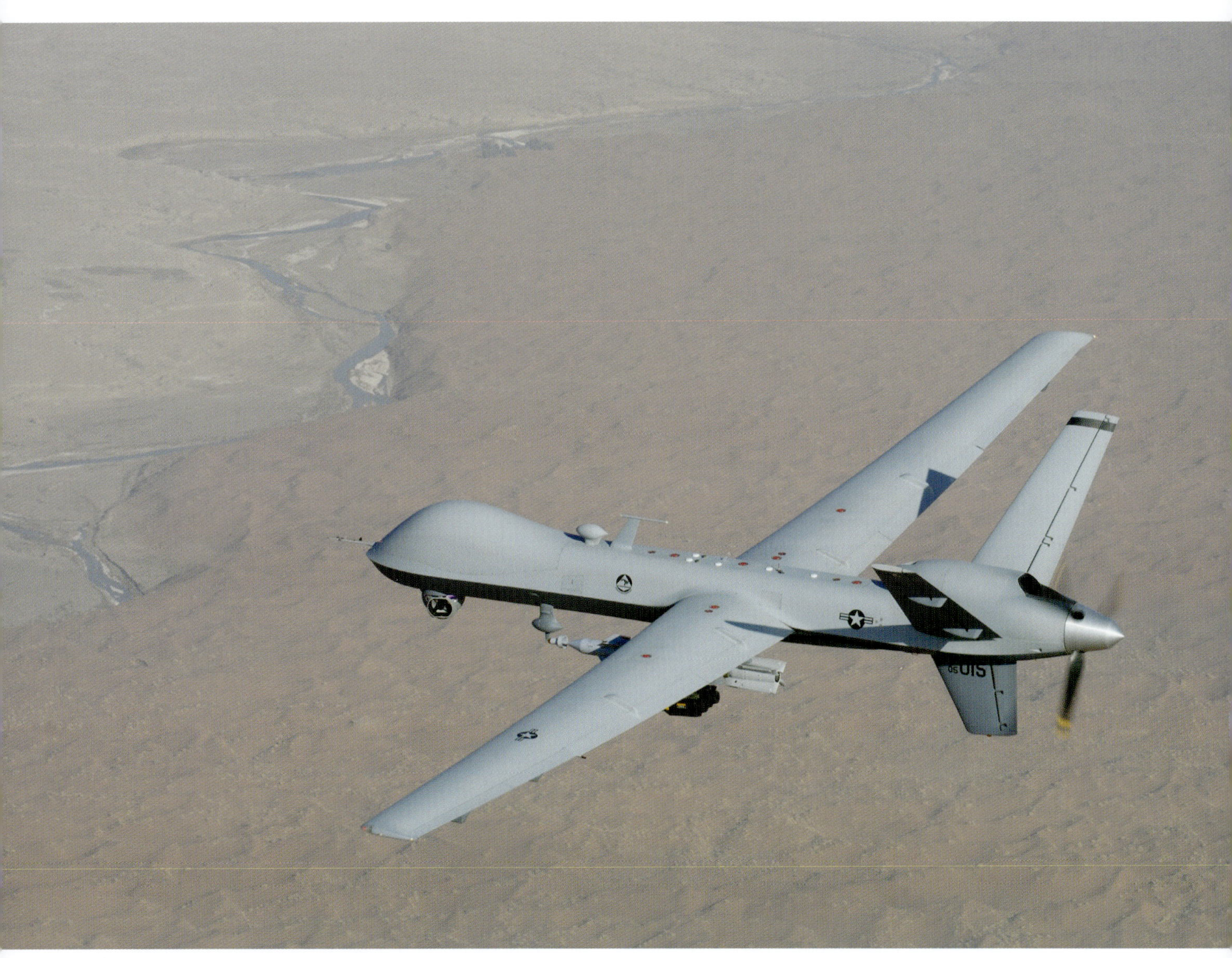

▲ A weaponized Reaper drone. The use of such machinery in recent conflicts has become increasingly controversial and has given negative connotations to the word "drone."

MULTIROTOR DRONES

Modern multirotor drones fly with three or more propellers and motors, and provide a stable platform on which to mount camera equipment. Many early multirotor drones were built by enthusiasts and required a fair bit of technical know-how to assemble and fine tune. These days, however, there is a wide range of all-in-one camera drones available to suit almost any budget and requiring little specialist knowledge to operate (although a healthy dose of common sense is a must!).

Mechanically, a multirotor drone is very simple: it flies by varying the speed of each motor and hence the lift produced by each propeller. However, there's a lot more needed under the hood. Just to keep a drone hovering requires a complex microcomputer (known as the "flight controller") that constantly monitors the aircraft and varies the speed of each motor. The speed of each motor is also varied according to inputs from the pilot or instructions from a computer program. Modern multirotor drones are therefore very stable and relatively easy to operate (especially GPS-stabilized models).

The design does have its limitations, though, and one of the biggest is flight time. In contrast to a fixed-wing aircraft, most of the battery's energy is expended fighting gravity—it requires a lot of power just to keep the aircraft in the air. As a result, a multirotor drone with a relatively small camera might only be able to achieve a flight time of 20–25 minutes. Flight times get much lower when bigger cameras are fitted, so a large multirotor drone lifting a cinema camera might be expected to fly for only 8–15 minutes before it needs a new set of batteries.

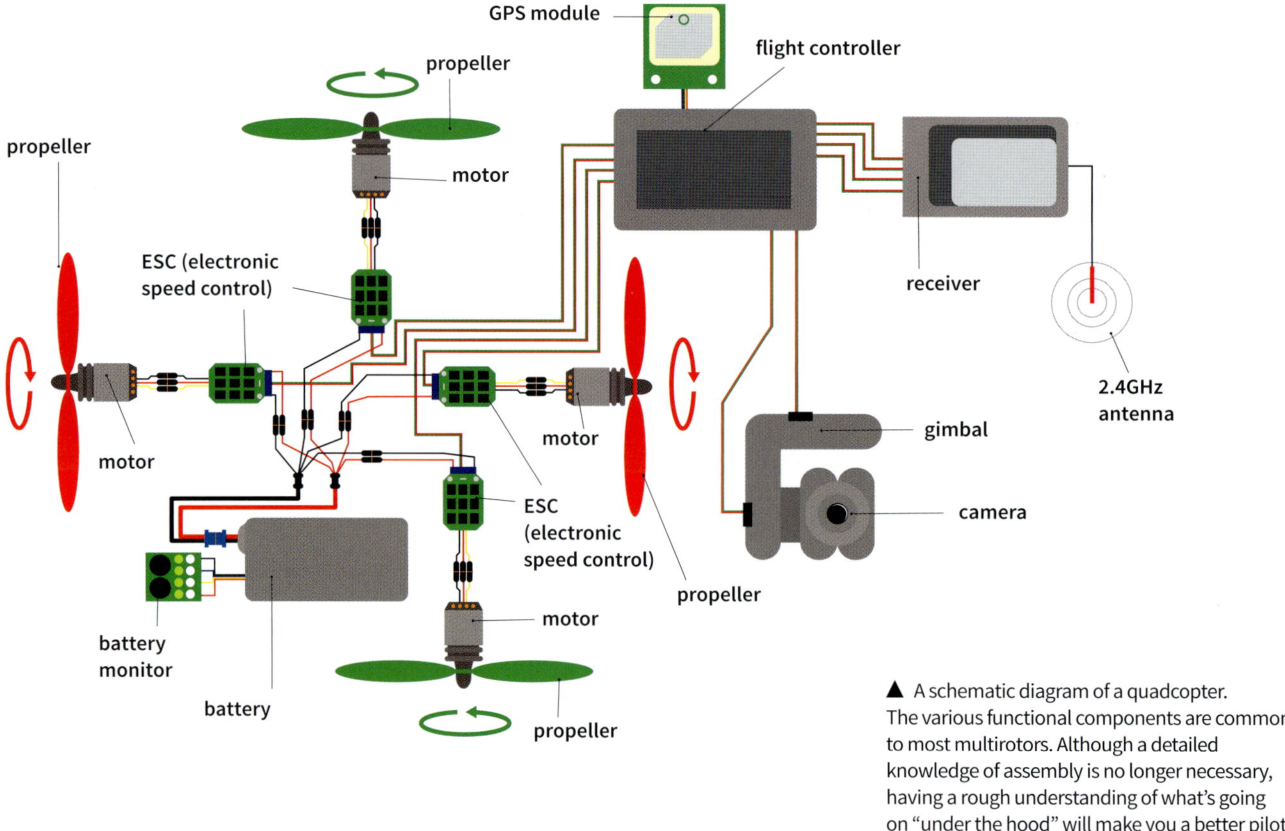

▲ A schematic diagram of a quadcopter. The various functional components are common to most multirotors. Although a detailed knowledge of assembly is no longer necessary, having a rough understanding of what's going on "under the hood" will make you a better pilot and improve your troubleshooting capabilities.

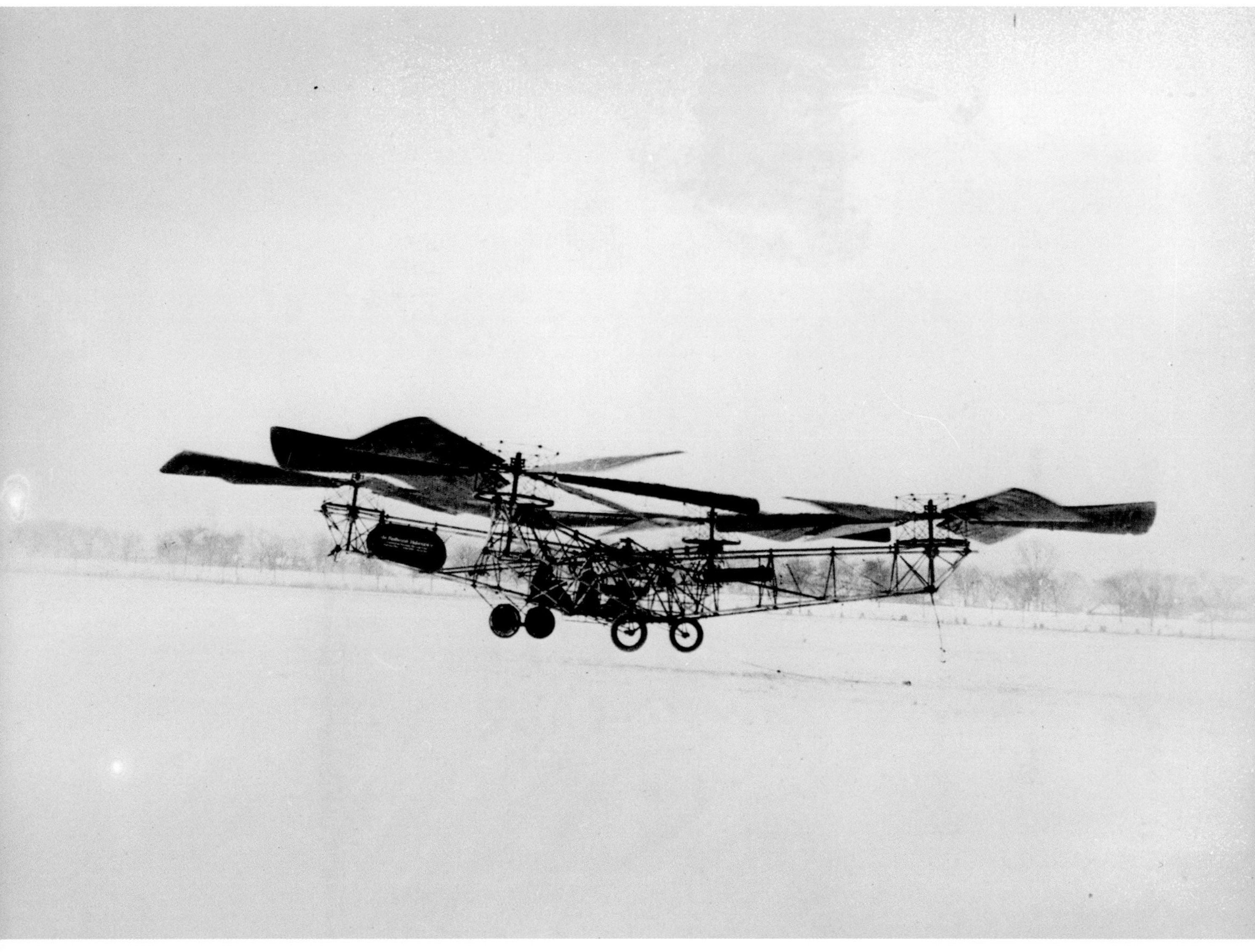

▲ The multirotor is not a new concept. This photograph was taken in January 1923, but manned multirotor aircraft are known to have been flying as early as 1907. However, the idea didn't catch on. Without the computerized control that we have today, there were major stability issues that made multirotors difficult to pilot.

FIXED-WING DRONES

Although not the main thrust of this book, it is also possible to attach a camera to a fixed-wing or "airplane-type" remote-controlled aircraft. A number of models is available, as outlined in the following chapter.

RAPID EVOLUTION

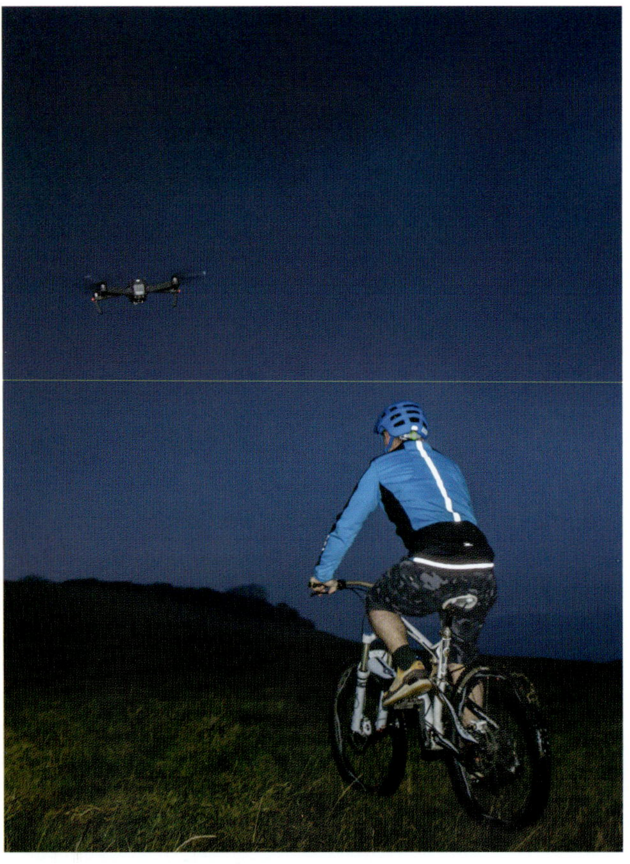

▲ Outdoor sports enthusiasts are embracing the portability inherent in the latest generation of camera drones.

▶ Camera-equipped drones have revolutionized the way you can photograph the world around you. No longer are you limited to where you can stand—the space above your head is also a potential camera position that can be used to provide you with fresh angles.

One of the most exciting—and frustrating—aspects of drones is how fast they are developing and evolving. It seems barely a few months go by without another new model being announced that has more features, a better camera, longer flight time, more smart flight features, and a lower price.

The great thing is that most of this development is happening to enthusiast-level camera drones, with a mind-boggling amount of technology increasingly being built into very compact drones. Technology from other sectors has also worked its way into modern drones, such as GPS navigation (originally used in ships and later cars), accelerometers (as found in cellphones), high-quality, miniature cameras (also pioneered in cellphones), ever more efficient Lithium Polymer batteries (a technology already in use in the radio-controlled world), and brushless gimbals (developed from handheld camera-stabilizing devices).

You can now buy a folding drone that will sit comfortably on the palm of your hand, is stabilized by GPS when outdoors, can avoid obstacles, has optical and acoustic sensors for flying indoors, can track moving objects automatically, be pre-programmed to fly to waypoints, and capable of shooting both high-quality 4K video and still photographs in Raw format—all for less than the price of a mid-range DSLR camera.

DJI

There is a reasonably strong emphasis in this book on DJI products. This is simply because they are by far the most popular small- to mid-sized drones on the market. However, most of what applies to DJI's drones applies equally to models from other manufacturers: although the exact methodology or terminology might vary, the principles remain the same.

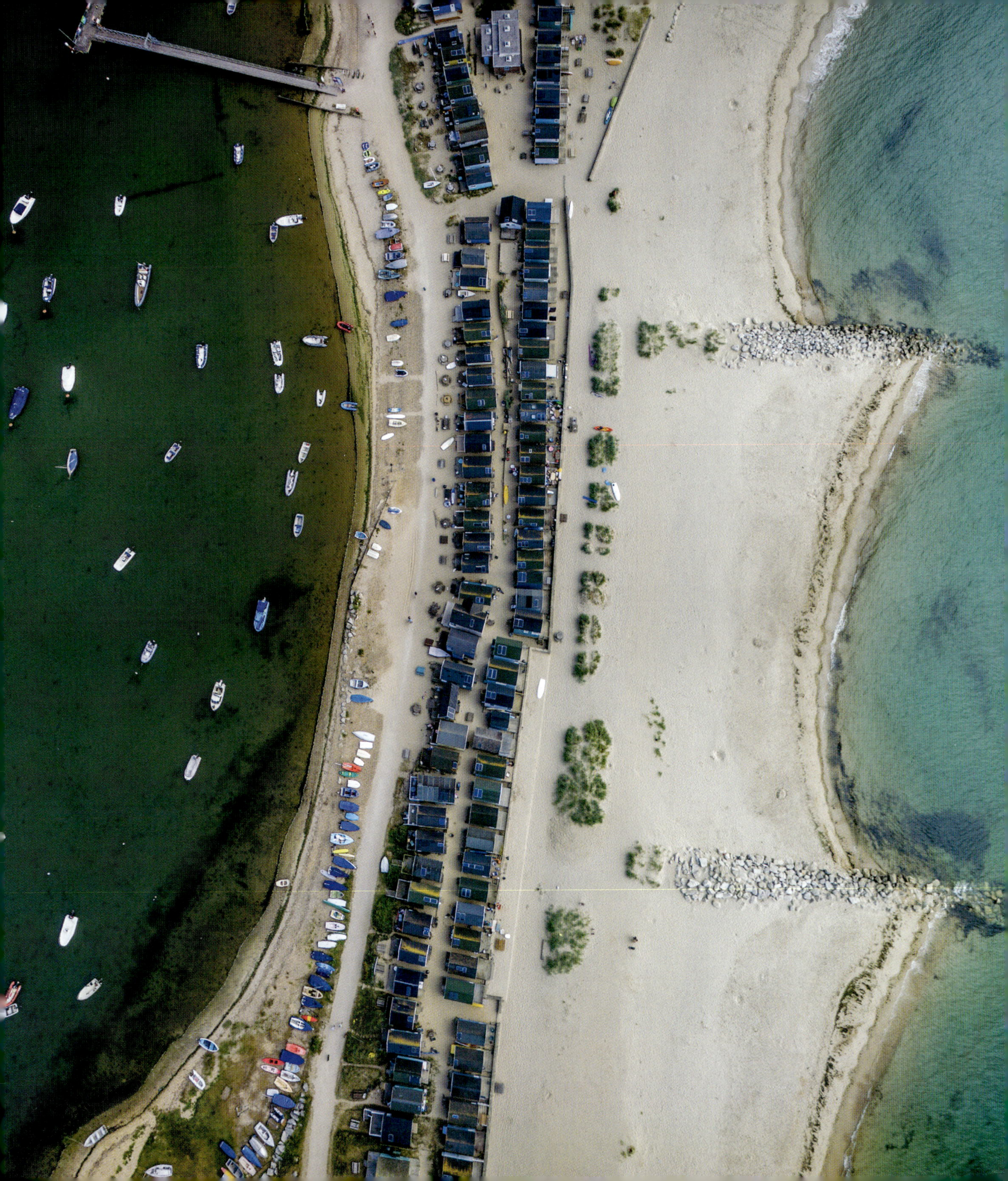

PROFILE: ANDY YEUNG

Andy Yeung is an award-winning photographer specializing in architecture and landscape photography. Based in Hong Kong, Andy uses the steep hills around the city as take off points for his drone. As well as being safer, this also takes him farther away from potential electromagnetic interference that is readily found in such a densely populated area.

Andy usually visualizes his photographs before he takes them, so he will know what lighting and weather conditions he needs. Although this will get him the results he wants, it also requires a great deal of patience—he may have to return to the same location multiple times over the course of months before the conditions create the composition he is after. This patience often pays off, though, as Andy's work has been recognized internationally, winning Gold in the Architecture category of the 2015 PX3 Photo Competition among other awards.

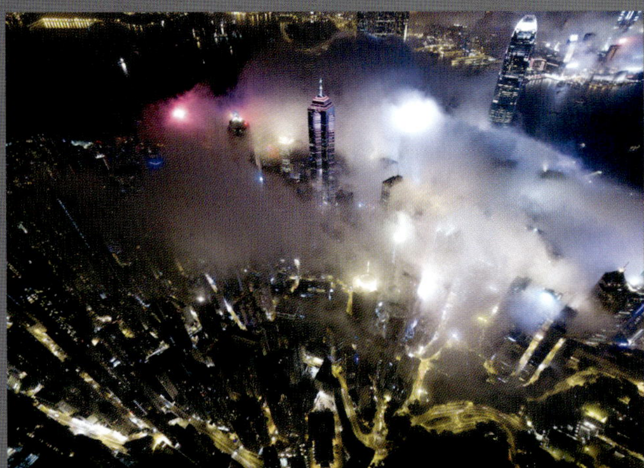

▲ Andy had seen a lot of amazing fog-themed photographs taken in the daytime and thought it would be interesting to capture Hong Kong in thick mist at night, from the sky. He loves how you can see the various colored lights of the buildings illuminating the fog from below.

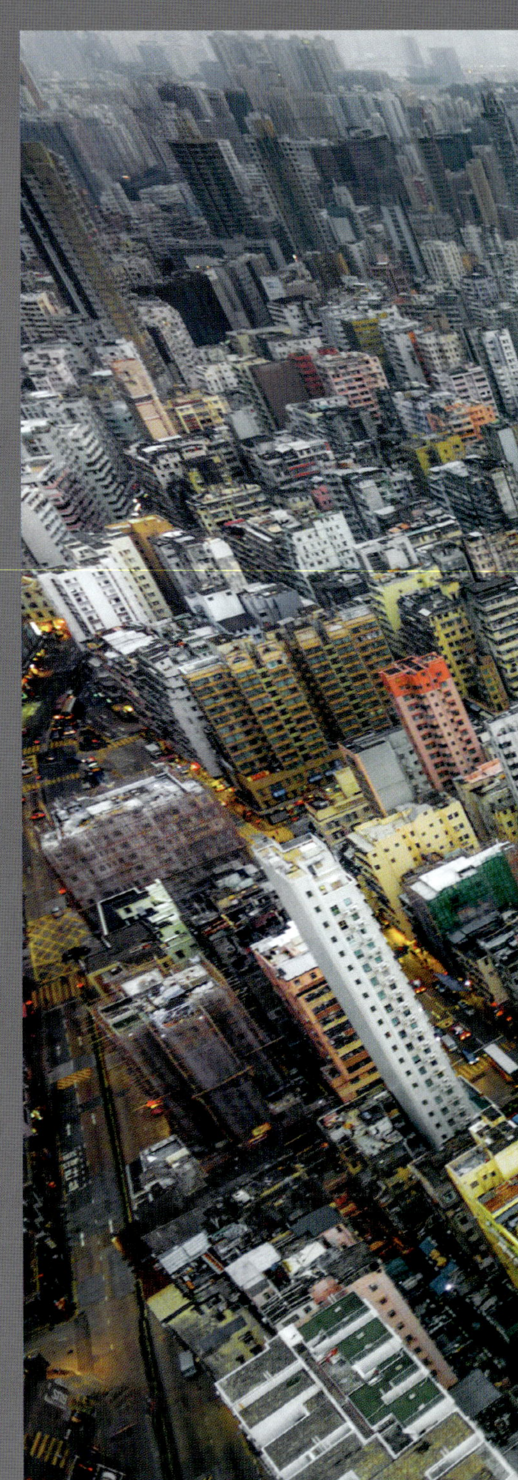

▶ Andy had the idea of creating a drone series showing Hong Kong from above when he was flying back from Europe and had an aerial view of the city. Looking down, he saw the buildings soaring high into the skies, offering a glimpse into the reality of living in one of the most densely populated areas of the world.

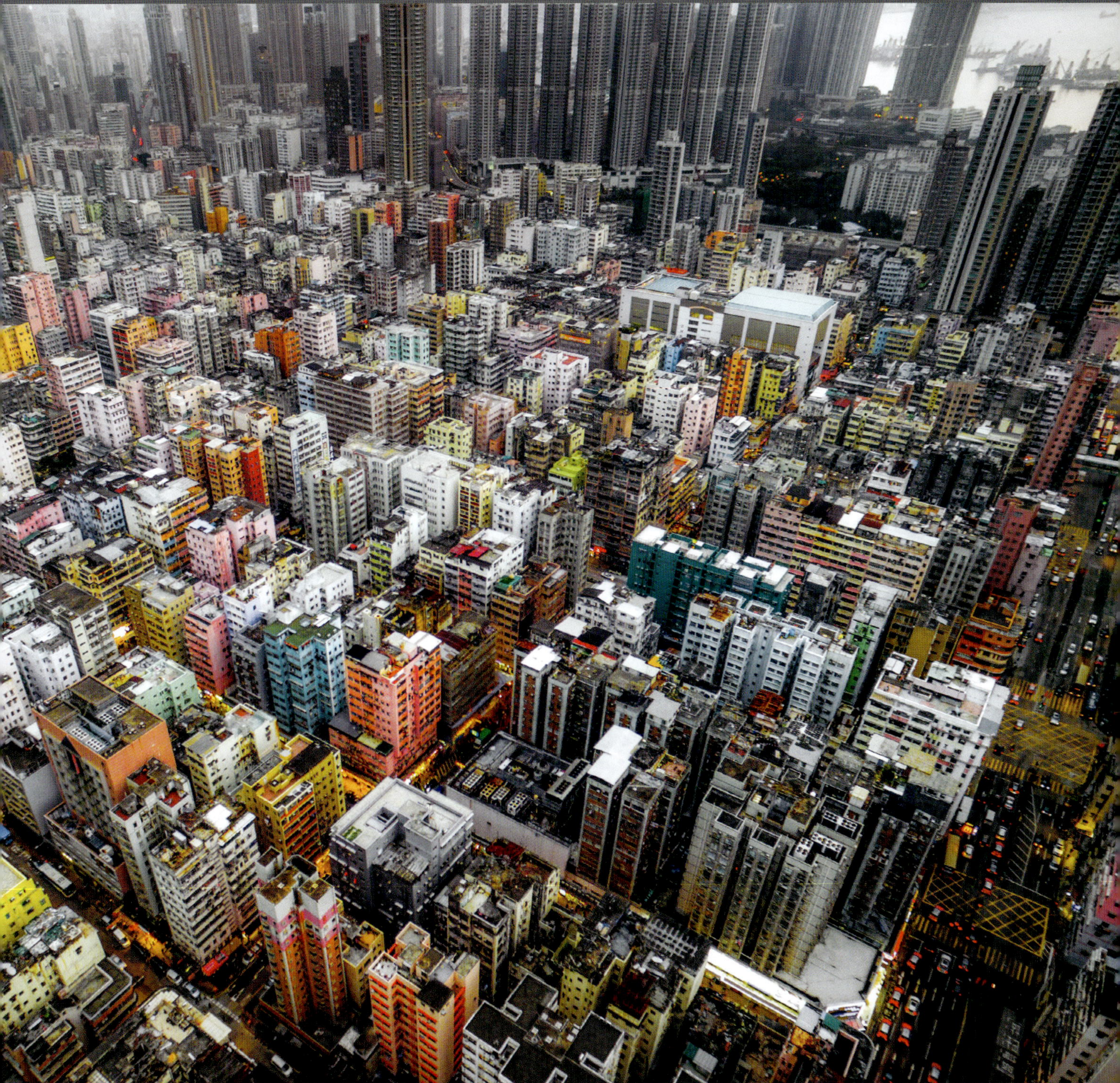

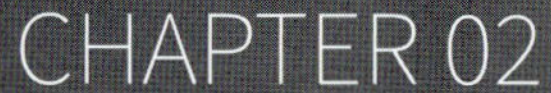

CHAPTER 02

EQUIPMENT

In the few years since drones first started to appear on the market, their development has been extremely rapid. It seems barely a month goes by without announcements of new and ground-breaking technologies. Generally, this is great, because potential buyers now have a lot of choice. It can also be quite confusing, though. In this chapter we'll explore the options, so you can ensure you pick the right drone for your needs.

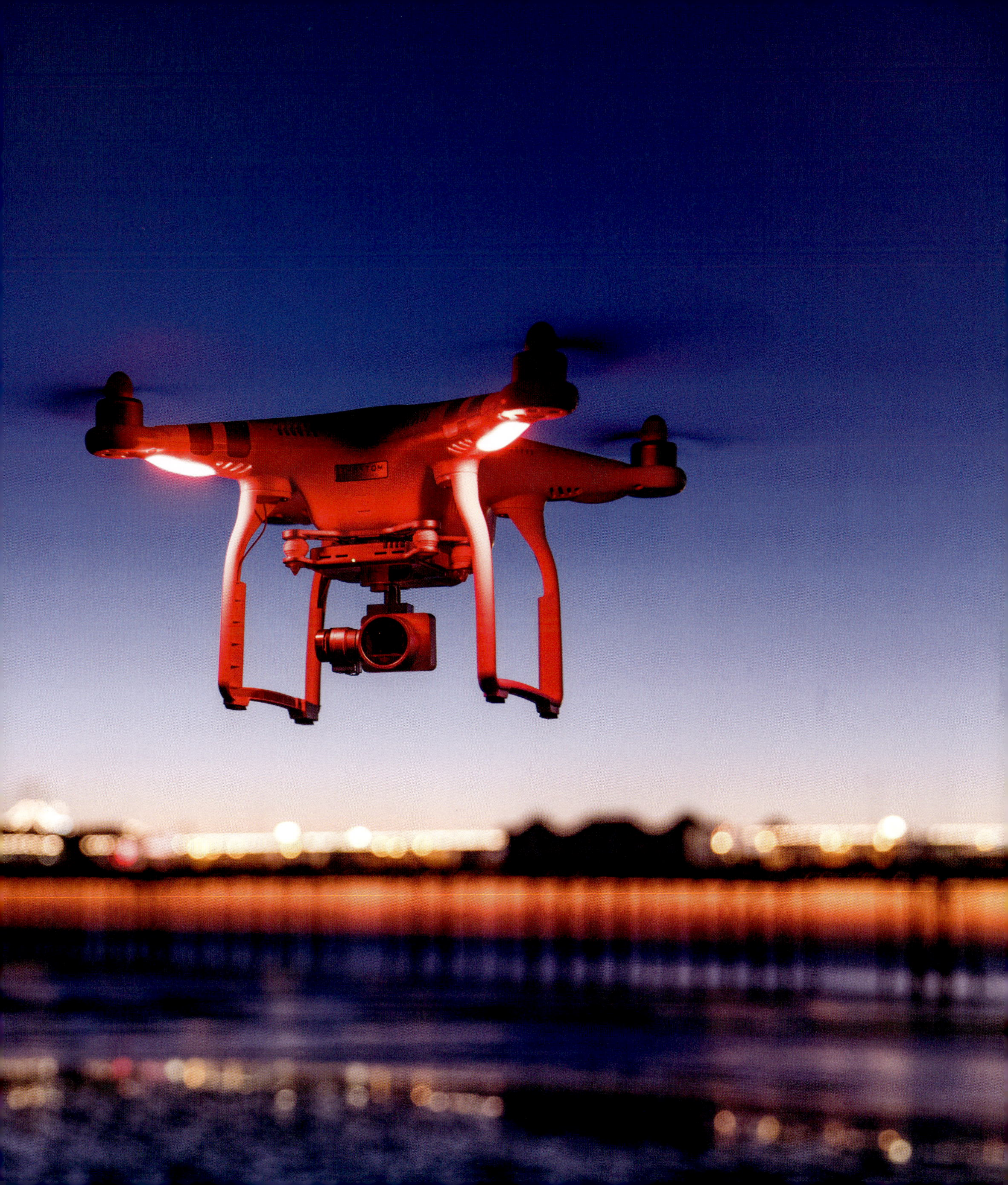

TYPES OF DRONE

In recent years the drone market has increasingly divided into more and more classes, with a wide variety of designs. The range of commercially available drones is categorized here, in approximate order of increasing size and cost. The list is by no means exhaustive—there are exceptions and models that straddle categories or defy categorization—but it should give you an idea of the current state of the drone market.

▼ A range of drones, from microdrones (front) up to a heavy-lift octocopter (back right). Be warned: drone photography can be both addictive and expensive!

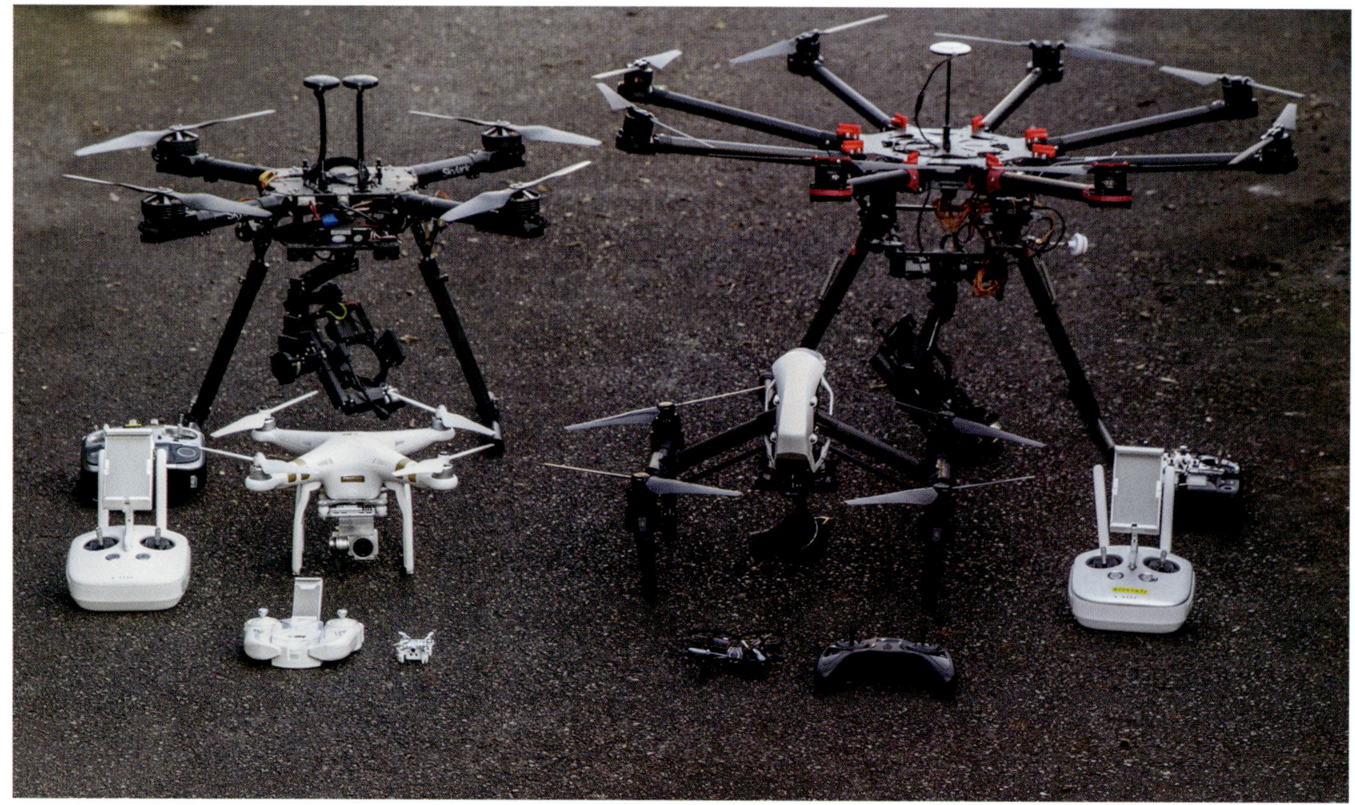

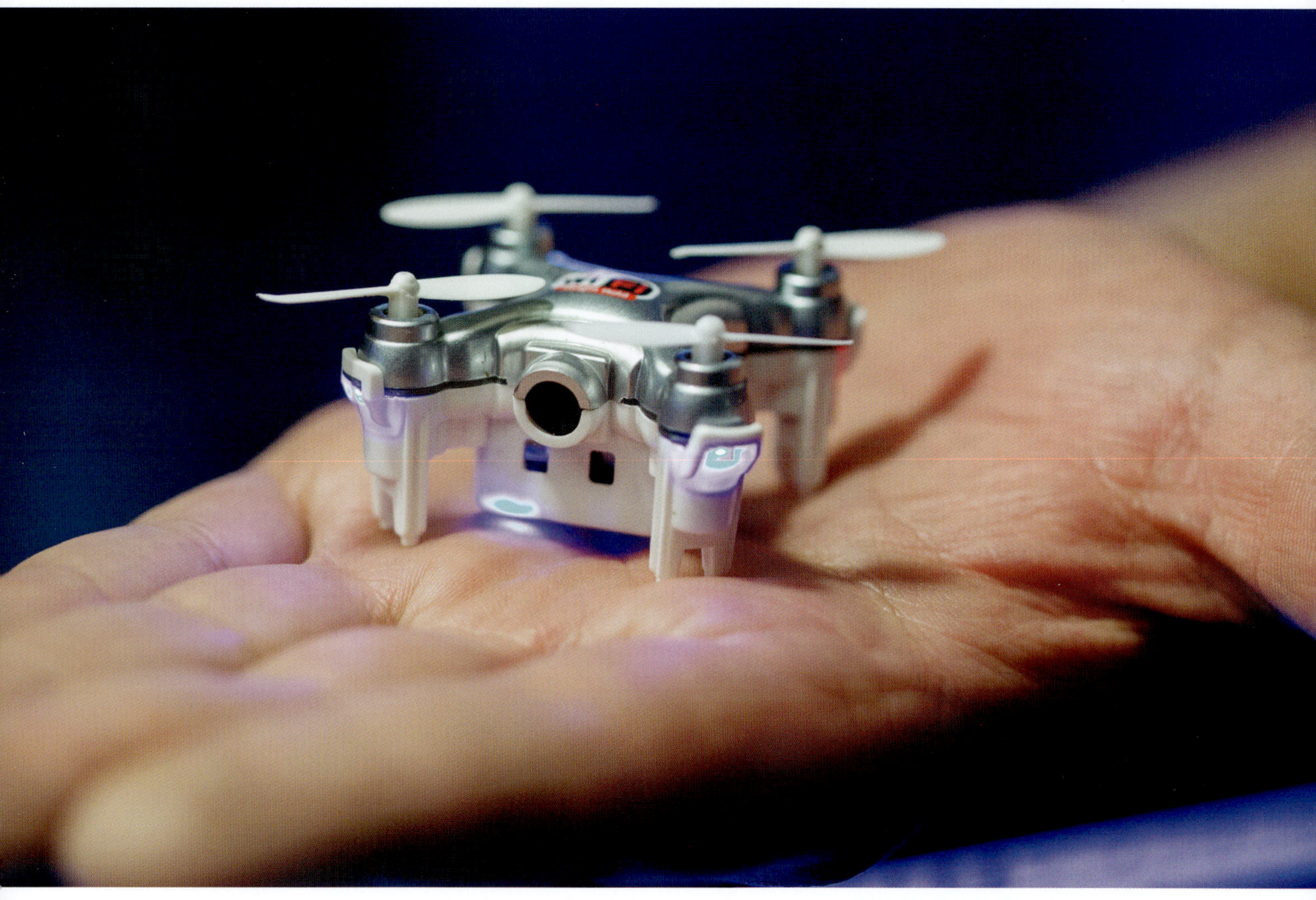

MICRODRONES

Microdrones typically measure 8in (20cm) or less in overall diameter and weigh very little. They are great for indoor use, but should be used with caution outdoors: they do not have GPS, will easily be blown around by the wind, and could be lost.

As microdrones are inexpensive they are a great way of practicing your skills before moving on to larger and more expensive models: their controllers work in the same way as larger drones, with a similar remote-control layout. Some microdrones are also equipped with cameras, but you should be aware that the cameras are tiny and aimed at the "fun" end of the spectrum, rather than serious photographic work.

▲ Microdrones are inexpensive, great for practicing with, and lots of fun to fly!

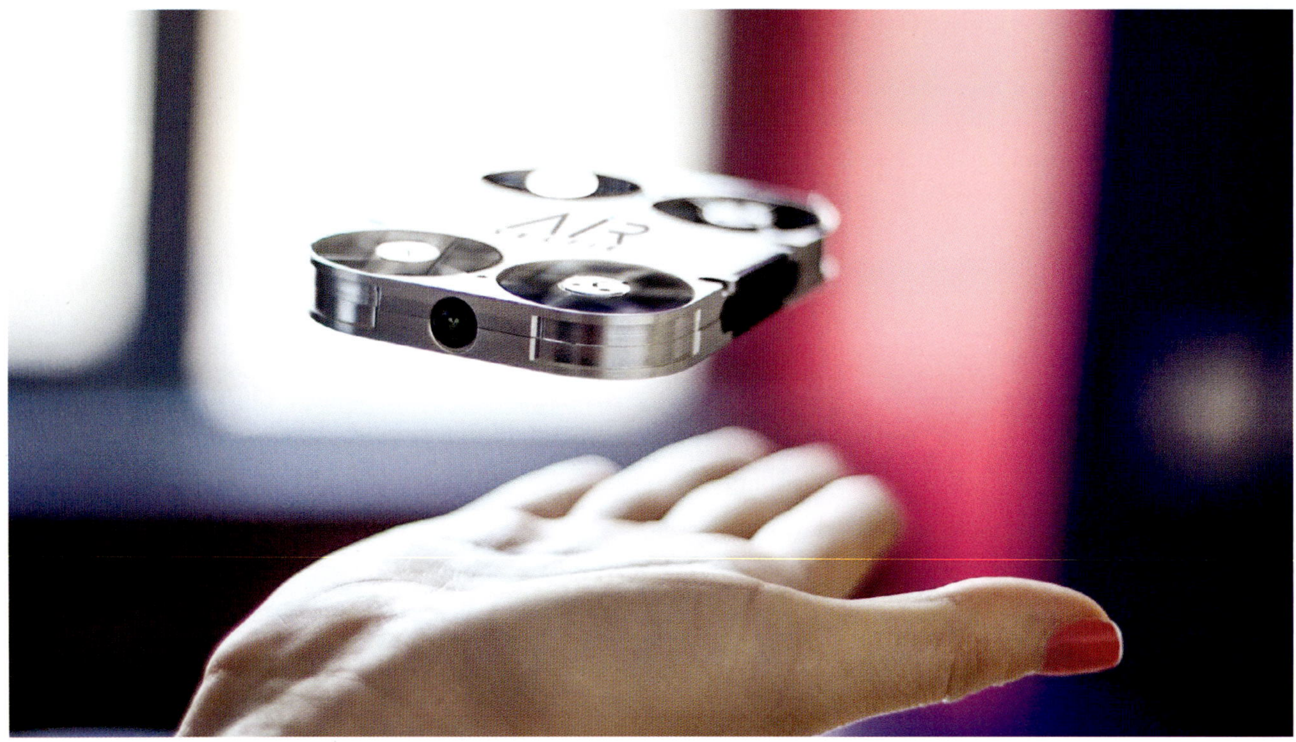

▲ The AirSelfie—the latest gadget for the social-media generation.

SELFIE DRONES

"Selfie drones" are a relatively new development and a step up from microdrones. They are typically pocket-sized folding designs with better cameras than microdrones. Very much aimed at the young, social-media savvy market, they are a fun way of getting some novel shots to put up on your Facebook or Instagram feed, but not really a serious photographic tool.

Selfie drones may have electronic image stabilization and—depending on the price—they may or may not have GPS stabilization. They are often controlled via a smartphone and may have a variety of automated flight patterns built in, such as flying up and away, keeping the user centered in the frame, or orbiting the user. Selfie drones are a fun gadget, but can be skipped by serious photographers.

RACING DRONES

These small quadcopters are designed primarily for speed, maneuverability, and durability. They are raced around a course of obstacles, usually several at a time: the action is fast paced and crashes are frequent!

As the pilots watch a live video feed from their drone through first person view (fpv) goggles, they are totally immersed in the race, and it is very reminiscent of a video game. The video feed is usually analog and comes from a small forward-facing camera for minimal latency (delay).

Drone racing is a lot of fun and it's a great way of practicing your flying skills, but this type of drone is not optimized as a photographic platform!

MID-SIZED CAMERA DRONES

This is the category that will be of most interest to prospective aerial photographers. Drones in this category will have a gimbal-stabilized camera capable of shooting good quality still photographs of 12 megapixels and higher (often with a Raw format option) and video up to 4K in resolution. The more expensive mid-sized camera drones have larger sensor cameras (typically 1in or Micro Four Thirds) and some have interchangeable lenses.

The larger drones of this type may also allow dual operators, so one person can control the drone and another can control the camera. These drones are equipped with a means of viewing the camera video feed, either via a smartphone or tablet connected to the controller, or on a monitor that is integrated into the controller. They may also have a variety of "smart" flight modes, such as orbiting a point of interest, following a subject, or waypoint-based navigation.

Mid-sized camera drones vary in size, from foldable models that will fit in a large pocket (or camera bag), up to aircraft in the region of 10–13lbs (5–6kg) that will require a large, suitcase-sized case to transport them in.

▼ DJI's Phantom range is the most popular mid-sized camera drone option on the market. They are easy to fly and have effective camera stabilization.

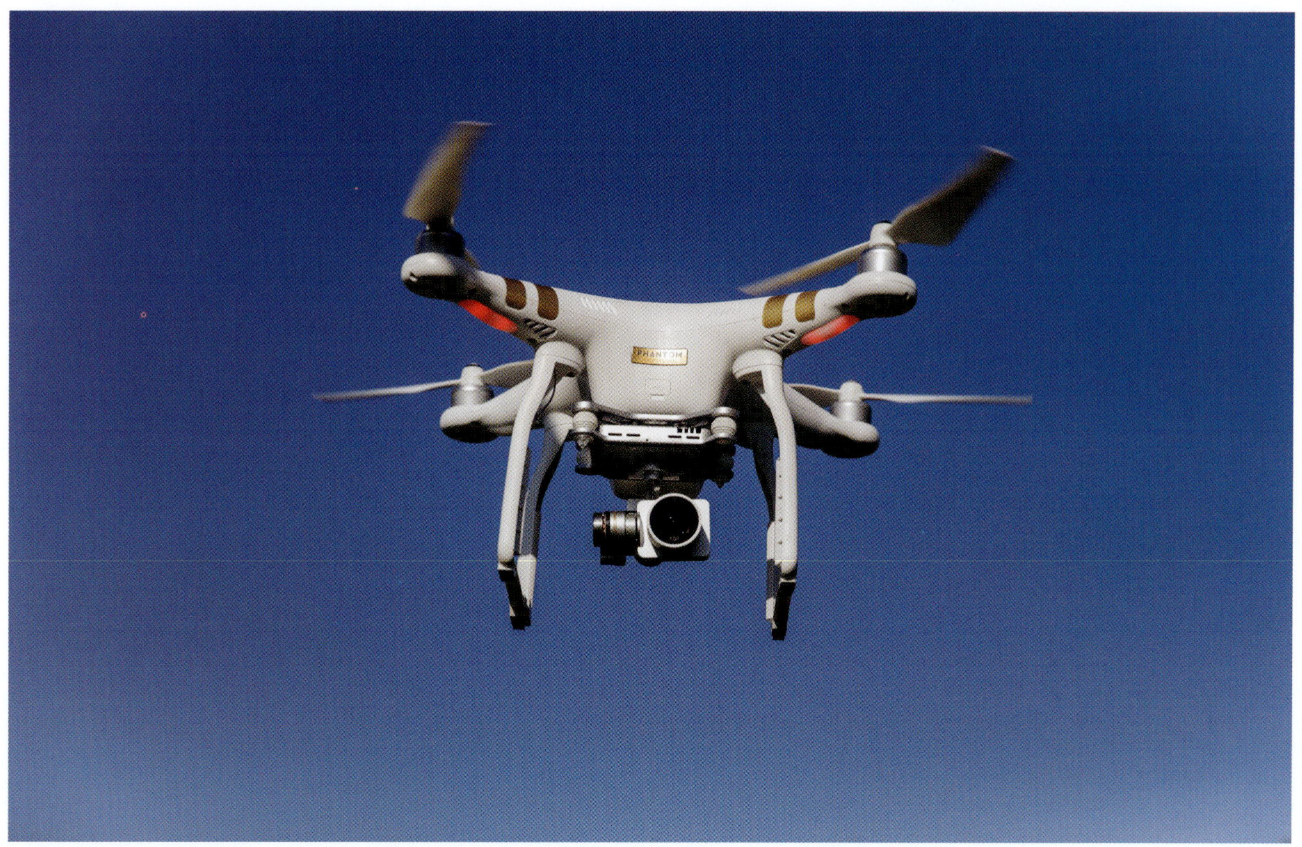

HEAVY LIFT MULTIROTORS

If you want to lift a camera the size of a digital SLR (or larger) then you need a heavy lift multirotor. These drones often require some degree of assembly by the operator, so they take considerably more time and effort to set up than a ready-to-fly, mid-sized multirotor. With more weight in the air you also need to be even more aware of any and all safety issues.

As well as supplying the camera, you will also need a large gimbal suitable for your model of camera: these gimbals will require care to set up with the correct balance and tuning.

Because of their complexity, heavy lift multirotors should only be taken on by those who already have considerable experience of multirotor flying. It is also worth noting that, depending on your exact setup, you may have less control over the camera than you do with an app-equipped mid-sized drone. However, for the ultimate in picture quality, these drones are currently the only choice.

▶ The large Aerigon drone from Intuitive Aerial can comfortably lift a cinema-grade camera such as a Red Epic or Arri Alexa Mini, complete with cinema lenses, and even a follow-focus (remote focusing device) if necessary.

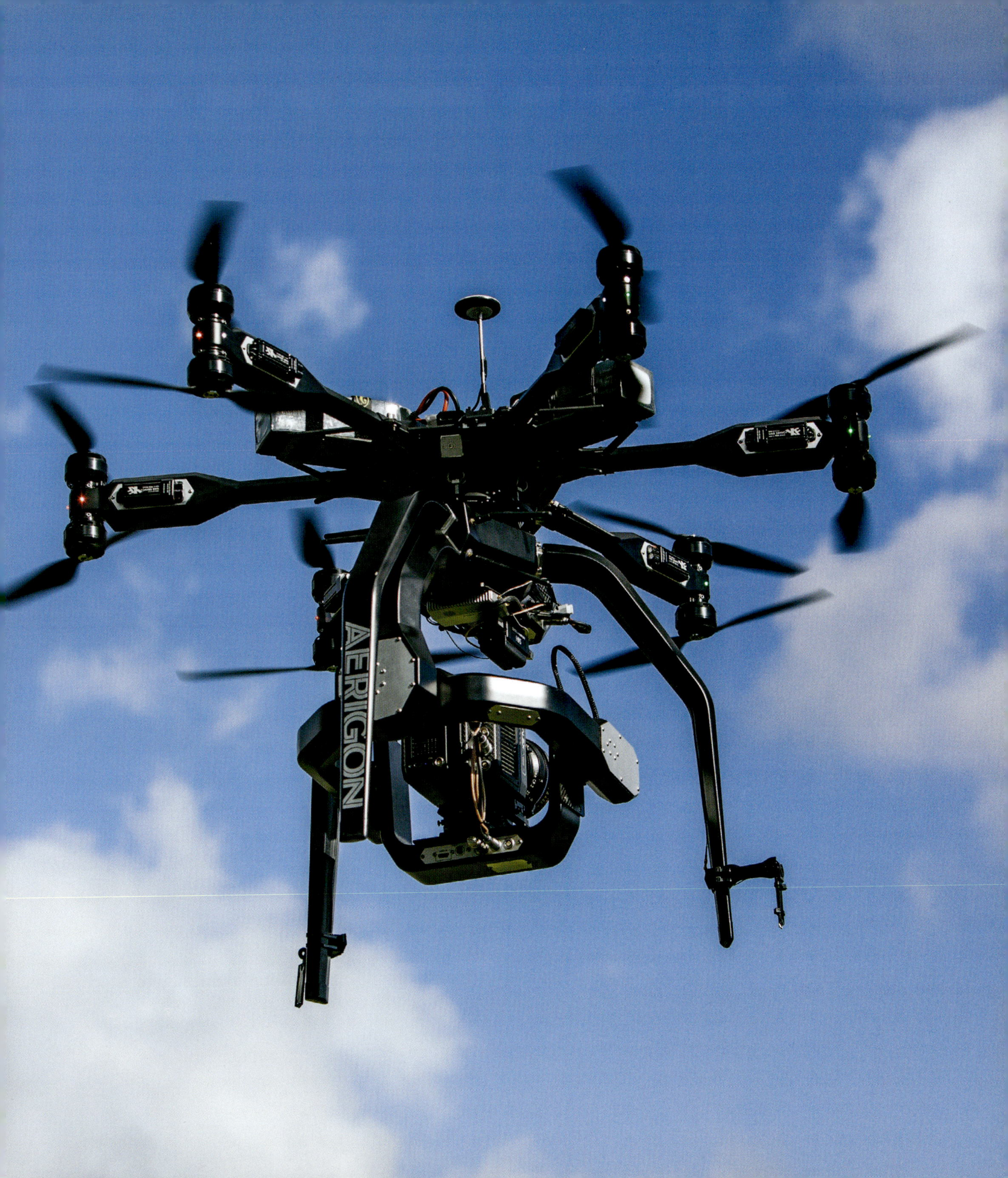
AERIGON

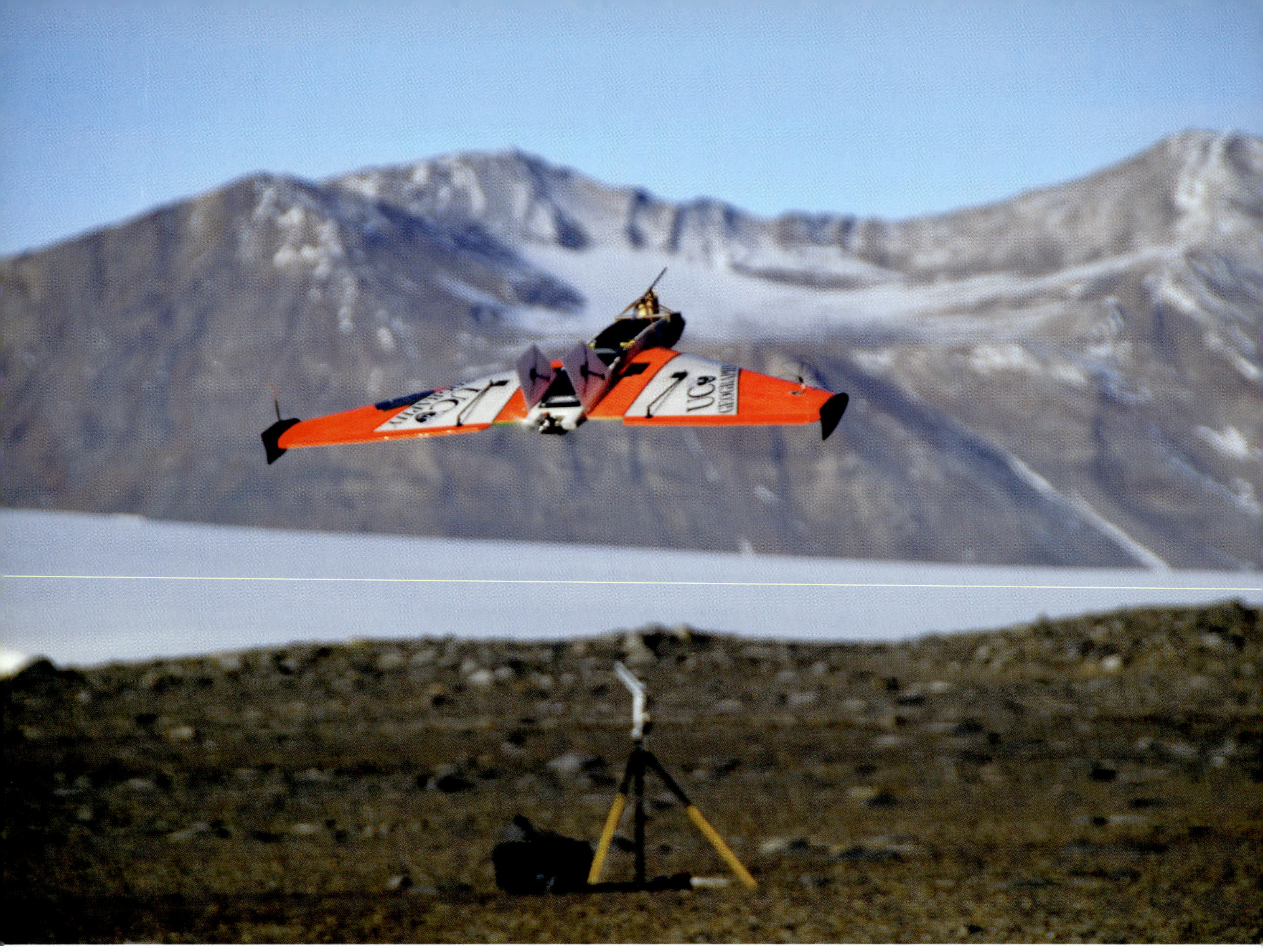

FIXED-WING DRONES

Fixed-wing camera drones cannot hover: they need to keep moving to maintain lift and stay aloft. For this reason flying them manually requires a lot more skill than flying a multirotor. Fortunately, many fixed-wing camera drones are at least partly computer controlled, making the task much easier.

In general, fixed-wing drones have a longer flight time than their multirotor counterparts, but they are not well suited to many conventional photographic tasks. However, they excel when it comes to automated mapping and surveying large areas, where flight time and speed are important.

One of the main disadvantages of fixed-wing drones is that they need a clear "runway" area for take off and landing. To address this issue, some fixed-wing drones have been developed with a vertical take off and landing (VTOL) capability, much like multirotor drones.

◀ Fixed-wing drones can fly quickly and stay aloft for long periods on a single battery. This makes them ideal for covering large areas during mapping or survey work.

▼ A Vanguard industrial surveillance drone. This class of drone will typically have a long battery life, weather-proofing, zoom lens, and/or thermal infrared capabilities. But this doesn't come cheap!

INDUSTRIAL & SURVEILLANCE DRONES

Surveillance drones often need endurance, weatherproofing, and a camera with a zoom lens. This industry is still in its infancy, but already there are large multirotor aircraft that can fly for over 90 minutes, cover a range of 20 miles (35km), come equipped with both a thermal infrared camera and a conventional camera, and can avoid obstacles autonomously. With these advances, Big Brother could be watching you from the sky very soon!

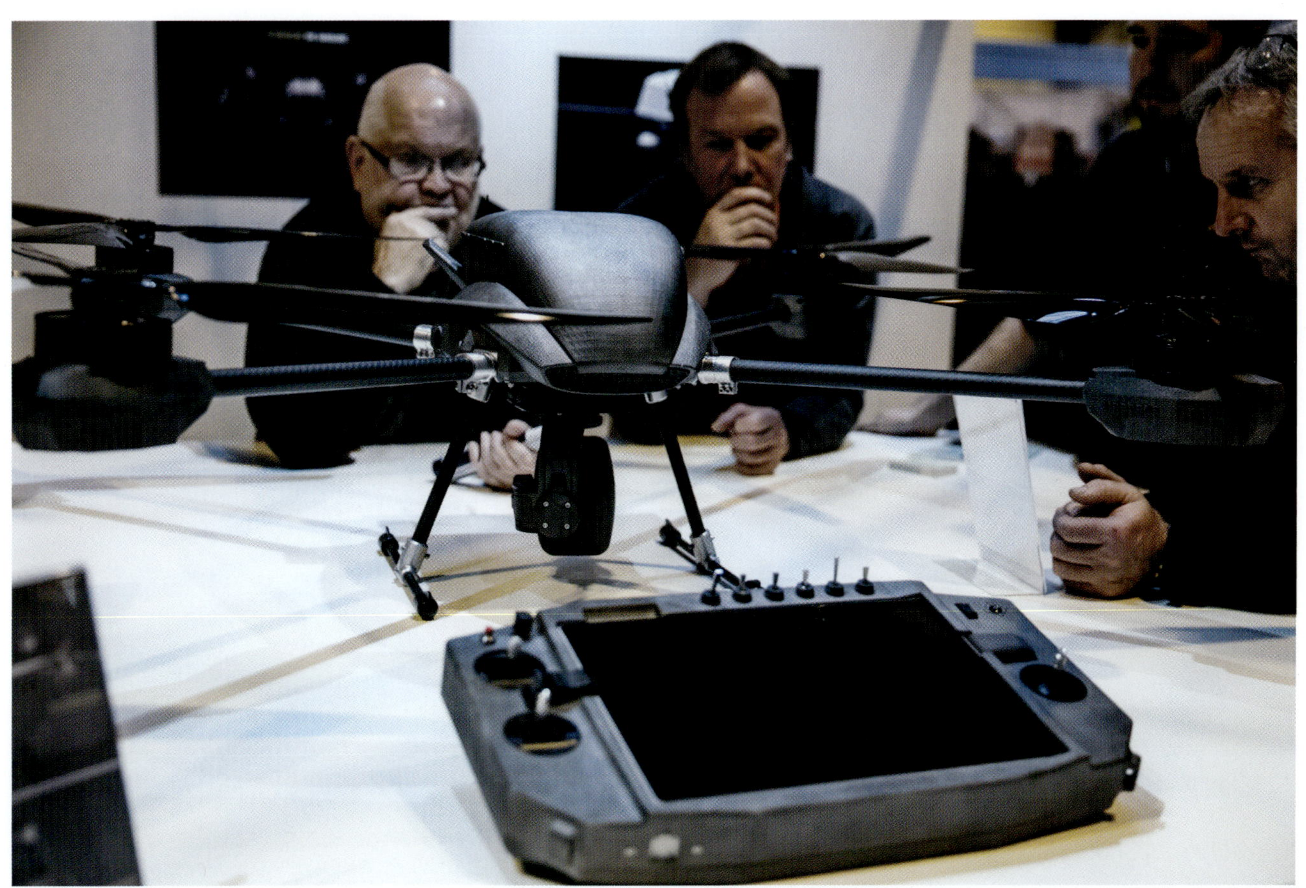

TRICOPTERS, OCTOCOPTERS & MORE!

Apart from fixed-wing drones, all drones are multirotor designs. However, there is a plethora of design layouts, ranging from tricopters (with three rotors) to large, heavy lift aircraft with 12 motors and rotors.

The optimum design layout is still hotly debated and there are many factors to take into consideration. For most consumer and semi-professional work, quadcopters (often shortened to "quad") seem to have become the most common layout. However, it can be argued that a hexacopter or octocopter design has more redundancy built it: if one rotor breaks or a motor fails, the drone has a reasonable chance of staying in the air (although this is by no means guaranteed and it may take some good flying skills to pull off a safe landing!). The counter argument is that you're also more likely to have a rotor or motor failure if you have six or eight of them, rather than four!

THE X FACTOR

Coaxial rotor layouts—which may be referred to as X6, X8, or X12 designs—have two counter-rotating propellers and motors on the end of each rotor arm. One advantage of this layout is improved redundancy, as a failed motor or propeller should have little effect on the aircraft's stability. Consequently this design is often used on very heavy lift multirotors carrying large, expensive camera equipment.

▶ The Aerigon drone has two motors and two propellers on the end of each of its six arms. This gives redundancy in the event that one of the motors fails.

▼ One of the more unusual drone designs available is the Powervision Poweregg!

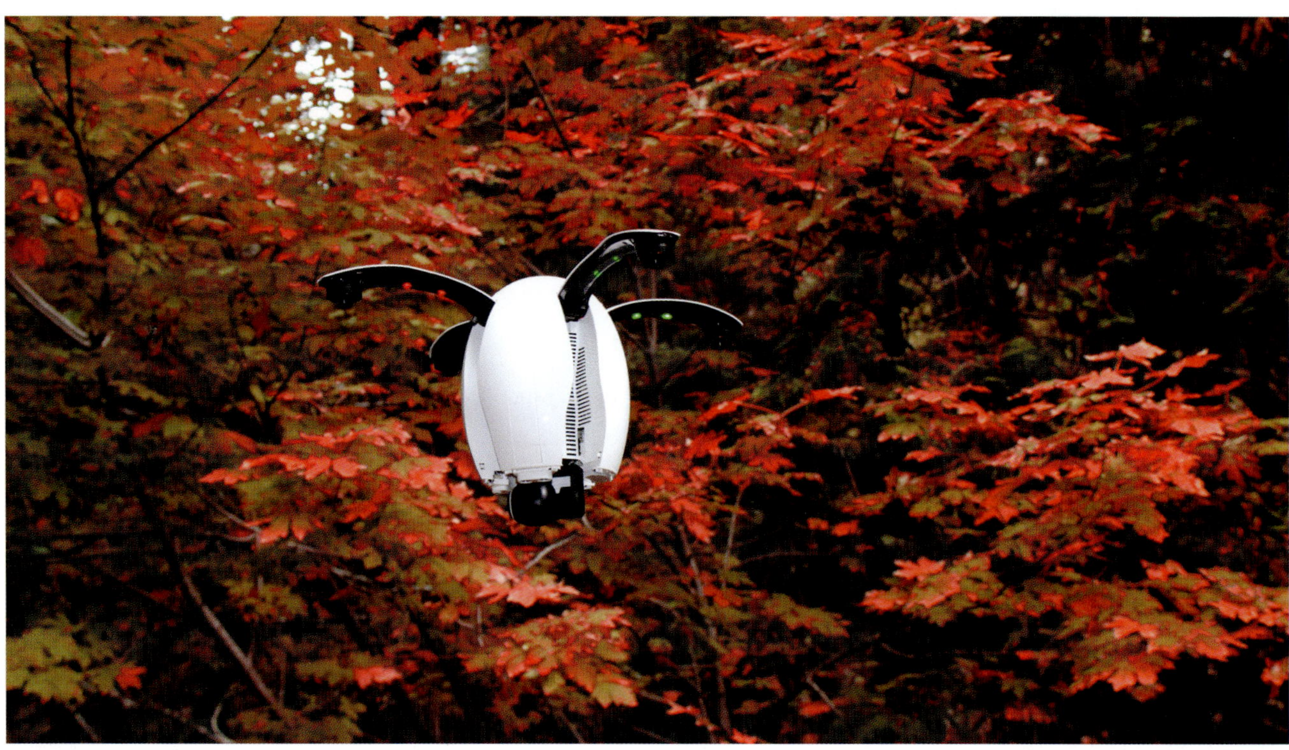

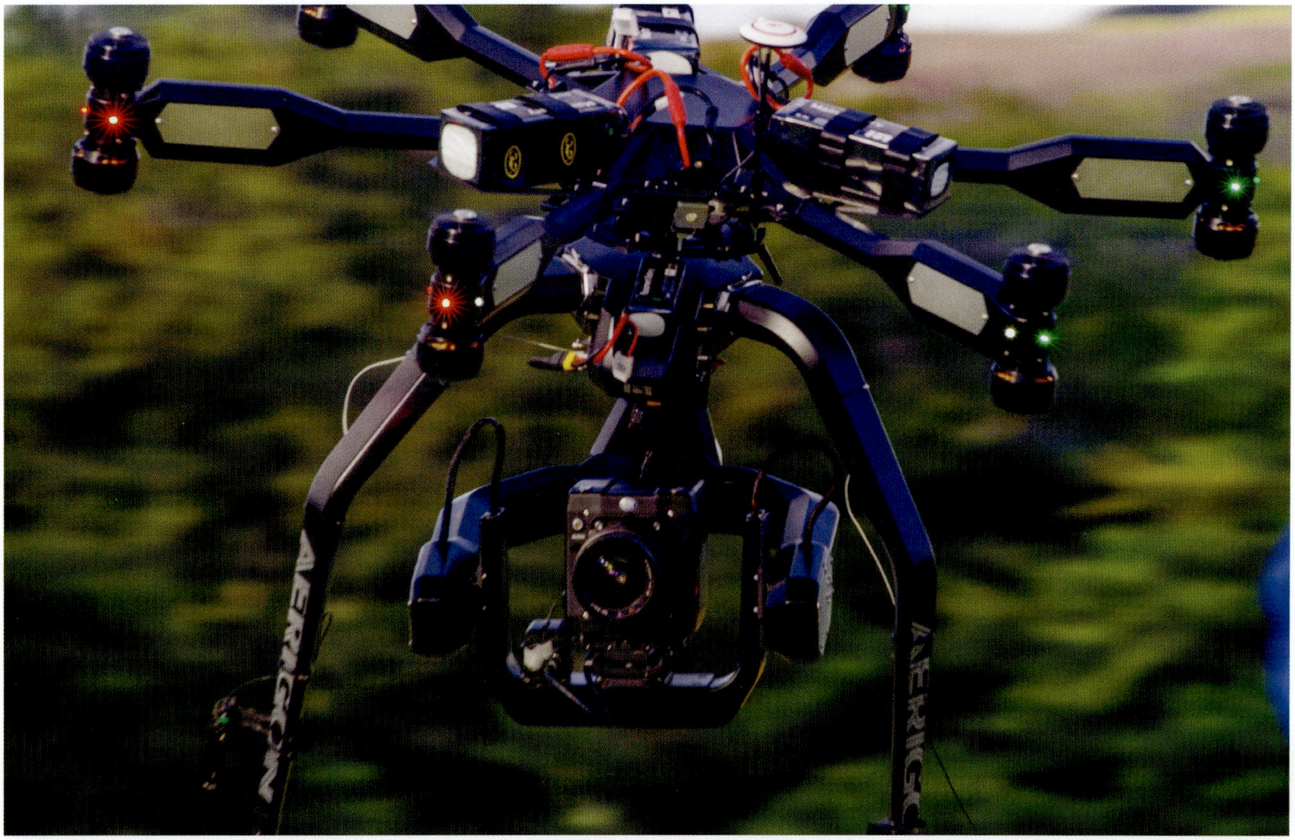

FOLD IT!

As high-quality cameras have been getting smaller and smaller, so have the drones that carry them. But even small- to mid-sized drones are a somewhat awkward shape to transport, especially if you have other gear with you. However, the latest folding drone designs allow you to pack a relatively high-performance camera drone in a very slim bag or alongside other camera gear in a regular camera bag. For some users this means the drone can be taken on a lot more photographic adventures than was previously feasible.

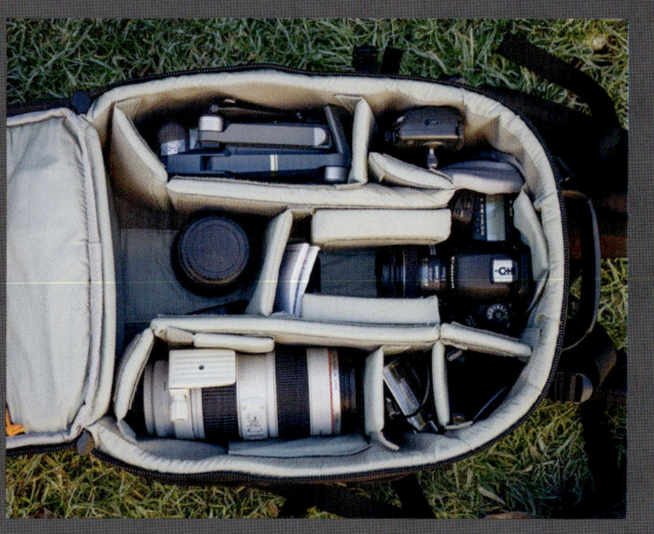

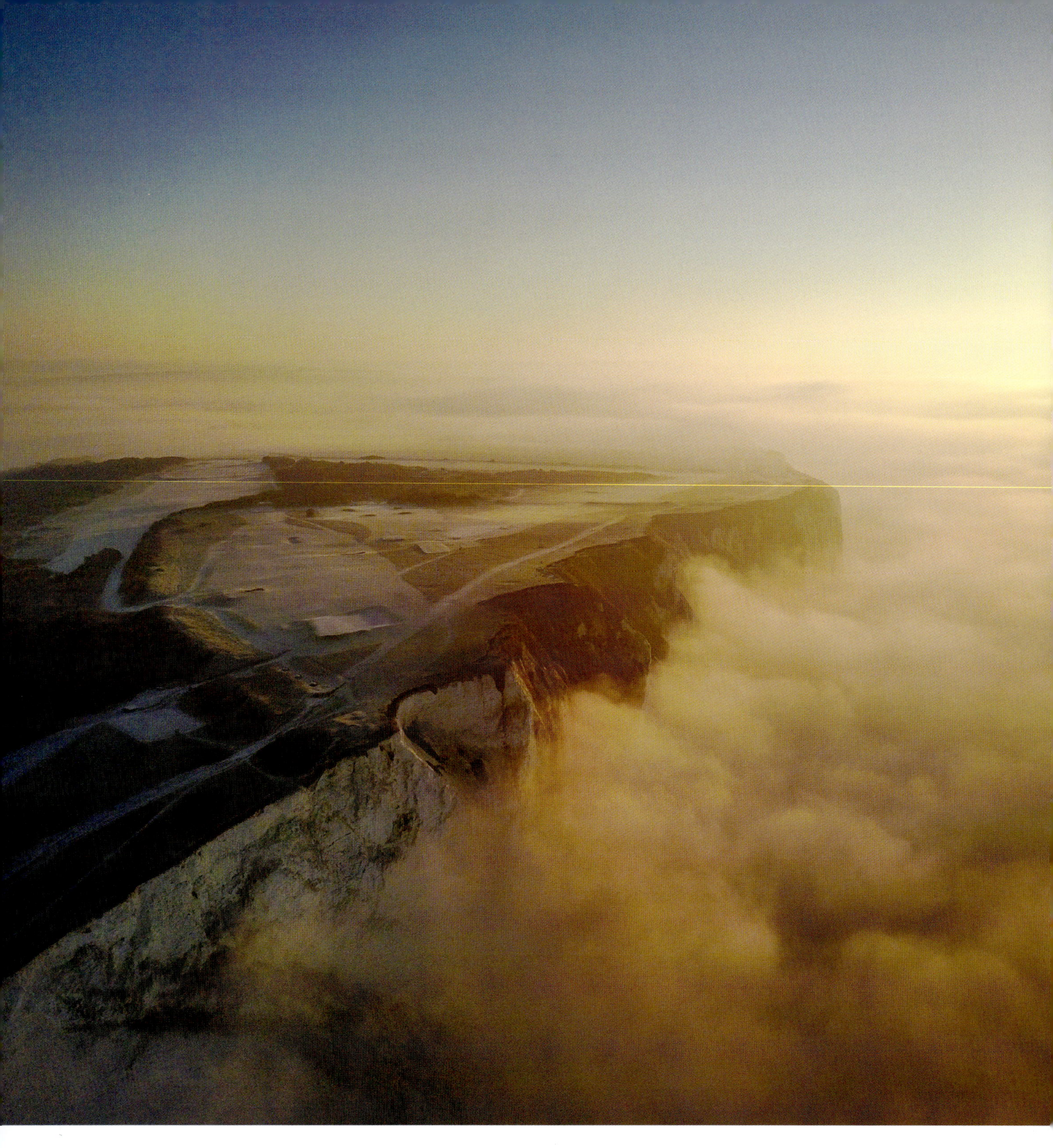

LANDING GEAR

On small- and medium-sized single operator drones, the camera is panned from side to side by yawing (twisting) the aircraft. This means the landing gear or legs will never get in the way of the shot, as the legs turn with the camera. However, on larger drones, the gimbal can rotate independently of the aircraft, so there is a risk that the landing gear *can* appear in view.

For this reason, many dual-operator systems have retracting landing gear, which allows the camera to rotate a full 360 degrees without obstruction. The pilot will typically raise and lower the landing gear with a switch on the remote, but there are also systems that can raise and lower the landing gear automatically in response to the drone's proximity to the ground. In practice I prefer to deactivate this feature, as I don't want a low-altitude video shot ruined by the sudden and unexpected appearance of the landing gear!

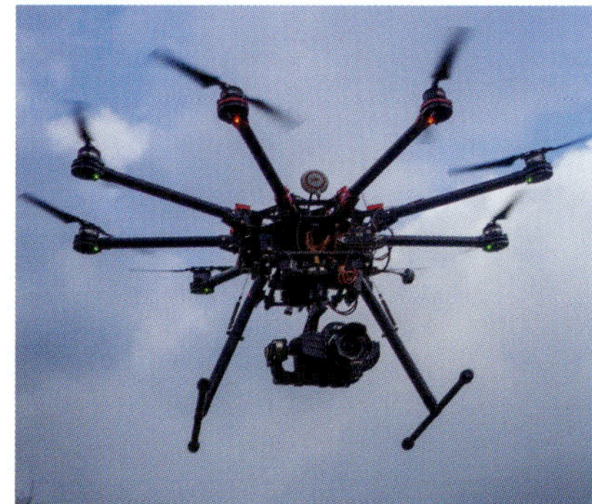

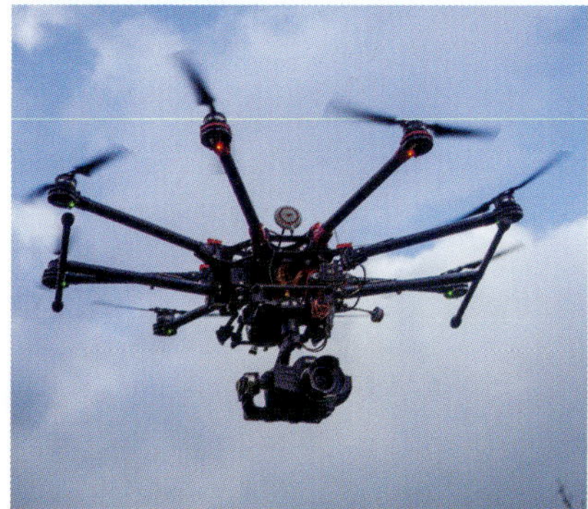

◀ Retracting landing gear is particularly useful when dual operators are shooting video. With the landing gear raised, the camera operator can follow the action through 360 degrees without obstruction.

▶ The retracting legs on this DJI S1000 lift out of the way to provide the camera with unobstructed 360-degree rotation.

GIMBALS

The simplest way of attaching a camera to a drone is via a solid mounting, and in the early days of drone photography this was all that was available. As long as the photographer uses a very fast shutter speed to avoid vibration-induced camera shake, and doesn't mind having very limited control over composition in flight, this works OK for still photographs. However, it's next to useless for video, as the constant vibration, wobbling, and tilting of the aircraft will result in nausea-inducing footage.

A certain amount of stabilization can be achieved inside the camera with sensor-shift and/or electronic stabilization (where there is effectively a moving crop of the full sensor), but for really effective stabilization the only solution is a motorized gimbal. Indeed, the advent of small, light, and affordable electronic camera gimbals revolutionized the quality of drone video.

An electronic gimbal is a platform for holding the camera with very precise electric motors controlling the movement of the camera on each of three axes—yaw (or pan), pitch (or tilt), and roll. Tiny and fiendishly clever accelerometers detect the movement of the drone in all axes and instruct the gimbal motors to compensate for the movement. The result is a rock-steady camera, even if the drone is wobbling all over the place.

The first really effective electronic gimbals available for the mass-market on small drones were for GoPro cameras, but they are now increasingly found in small- to medium-sized drones as well. For a large proportion of drone users this will be the norm and they will not need to worry about choosing a gimbal and camera/lens combination that works well: it's all been done at the design and manufacture stage. The most thought the user will need to put into the gimbal will be to calibrate it occasionally via an app and either use a gimbal lock or detach it entirely from the drone during transport.

However, many high-end professional users will demand to use larger or specialist cameras. If you need to buy your camera and gimbal separately from the drone, then it's very important that the two are compatible. In the case of interchangeable lens cameras you also need to check that the camera/lens combination will balance on your gimbal of choice.

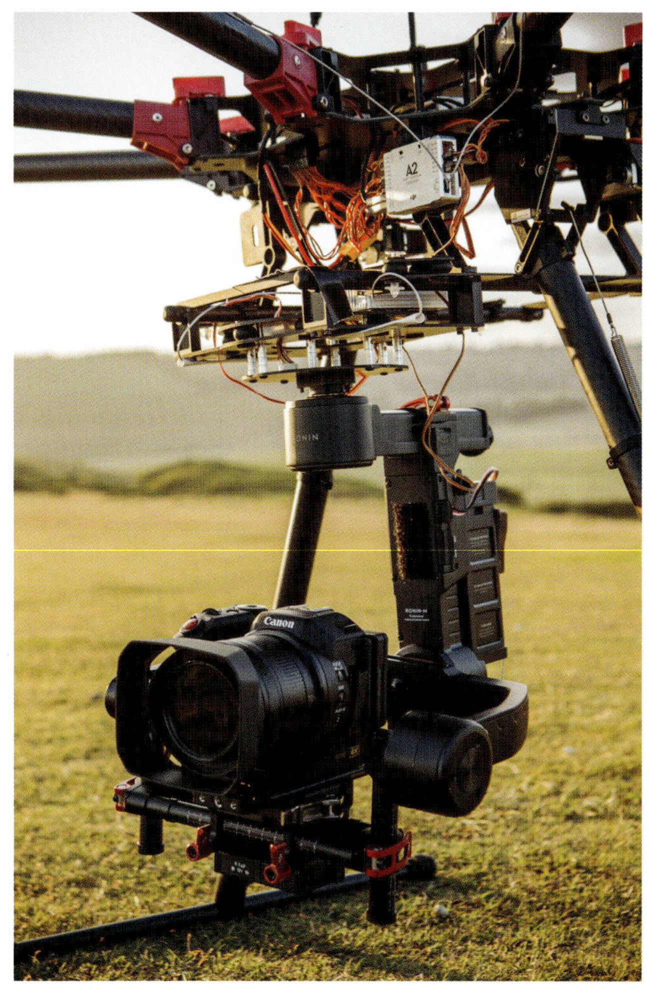

Broadly speaking, drone gimbals fall into two categories:

- Gimbals manufactured specifically for drone use that are tailored to one model of camera (and often one lens model).
- Gimbals adapted from handheld models that can be used with camera/lens combinations within a certain weight range.

The former group is generally much quicker to setup, but is more limited in scope. The latter group is far more versatile, but inevitably more time will be spent balancing the rig, calibrating the gimbal, and setting up remote gimbal control, gimbal power, and HDMI convertors (amongst other things).

◀ An adapted handheld camera gimbal. These handheld stabilizers can support a range of different cameras, but take considerably more time to set up than a dedicated integrated aerial system and may incur the weight penalty of an additional battery.

▼ By dispensing with the rear LCD, battery, and buttons, a small integrated camera, as found on the DJI Mavic (below) can be made very compact and light. However, standalone cameras, such as a full-size DSLR (bottom) still set the standard for image quality.

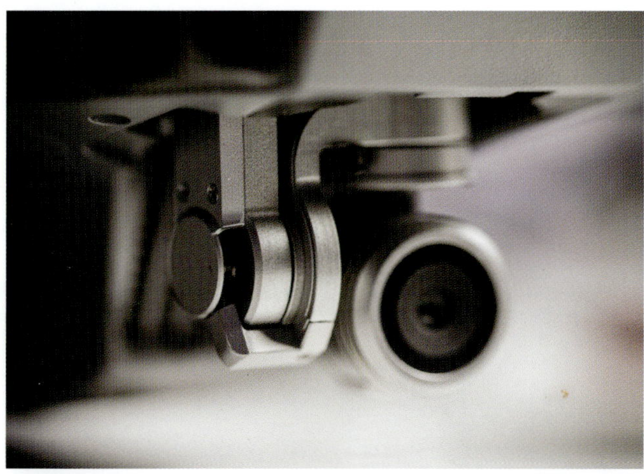

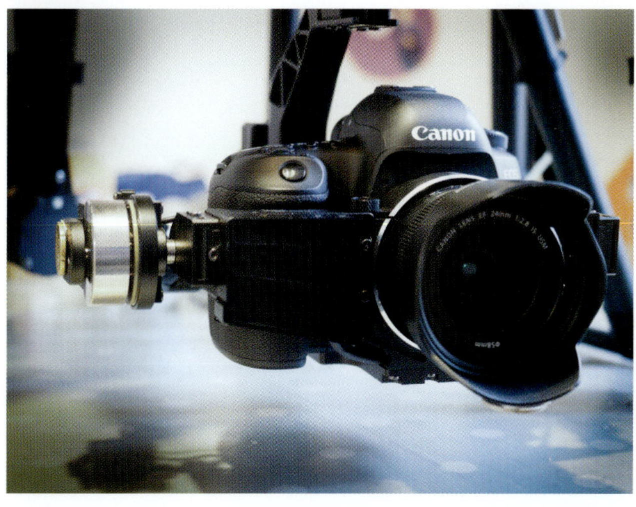

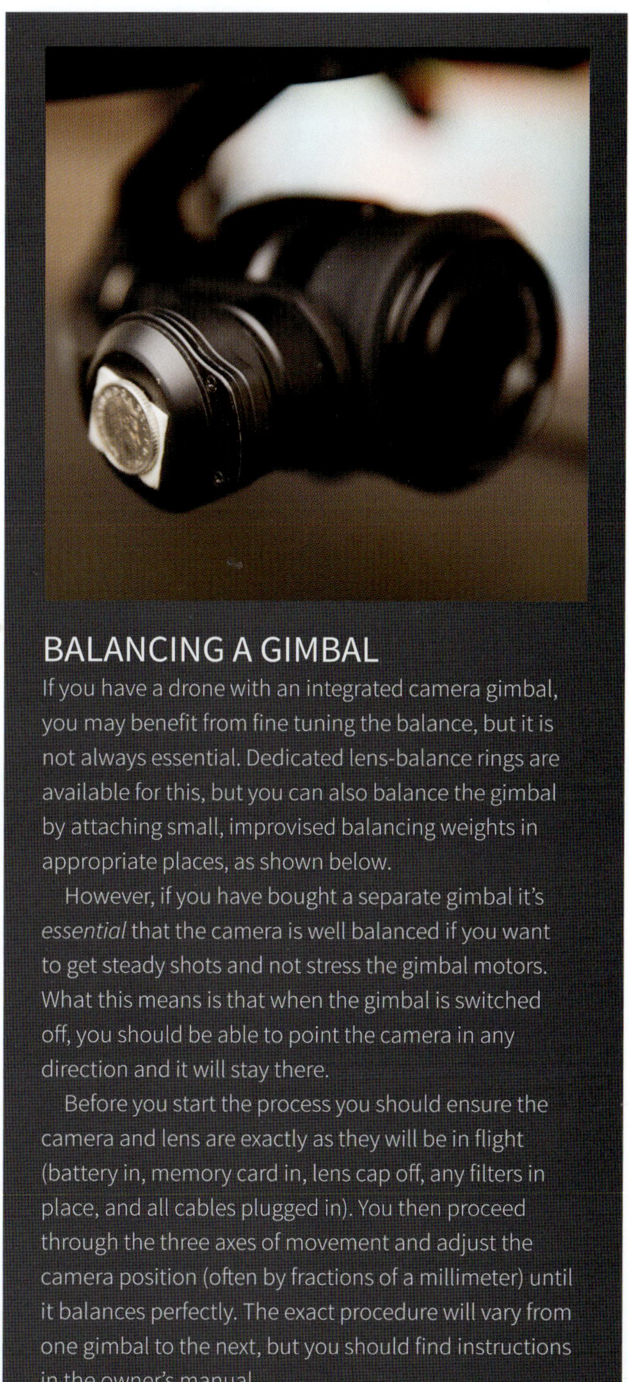

BALANCING A GIMBAL

If you have a drone with an integrated camera gimbal, you may benefit from fine tuning the balance, but it is not always essential. Dedicated lens-balance rings are available for this, but you can also balance the gimbal by attaching small, improvised balancing weights in appropriate places, as shown below.

However, if you have bought a separate gimbal it's *essential* that the camera is well balanced if you want to get steady shots and not stress the gimbal motors. What this means is that when the gimbal is switched off, you should be able to point the camera in any direction and it will stay there.

Before you start the process you should ensure the camera and lens are exactly as they will be in flight (battery in, memory card in, lens cap off, any filters in place, and all cables plugged in). You then proceed through the three axes of movement and adjust the camera position (often by fractions of a millimeter) until it balances perfectly. The exact procedure will vary from one gimbal to the next, but you should find instructions in the owner's manual.

CAMERA CHOICE

Unlike a ground-based camera, the weight of an aerial camera is often a major consideration, as every additional ounce or gram incurs a flight-time penalty. This is one of the reasons why drone manufactures are increasingly producing aircraft with integrated cameras: if the camera does not need to work independently of the aircraft, features such as a rear LCD screen, handgrip, and a plethora of buttons and dials can be removed (even the battery can be discarded as the camera can draw power from the aircraft's flight battery). This makes a more efficient setup overall, with relatively small drones capable of lifting high-performance cameras.

These integrated cameras usually have a relatively small sensor (around ½in) and a fixed focus, wide-angle lens (often delivering a horizontal field of view of 70–90 degrees). They are great general-purpose cameras, but the image quality will decline markedly in low light, due to the small sensor. The video is also compressed quite heavily (low bitrate), meaning that under some circumstances fine detail can be lost, especially if you later need to carry out extensive post-processing.

Many integrated cameras have the ability to shoot Raw still images as well as the more familiar compressed JPEG format, which is a welcome feature for image-quality aficionados. However, there will always be a minority of users whose applications are even more demanding when it comes to camera specifications. They might need very high-resolution still images for large prints, better low-light performance, and perhaps interchangeable lenses. If you have these camera demands then you will usually need a significantly larger and more complicated drone and gimbal setup. Checking all the components are fully compatible in advance is essential.

The other thing to bear in mind when using a non drone-specific camera is that you may only have a limited ability to control the camera while the drone is in the air. Typically you can trigger the shutter or start/stop video via a channel on the remote, but other than that, you need to set the exposure, focus, and other vital settings on the ground.

It is also worth noting that stress increases with payload. I have flown drones with extremely expensive DSLR cameras and high-end cinema cameras, and even knowing that you are fully insured, the sheer value of equipment suspended in thin air adds considerably to the stress factor!

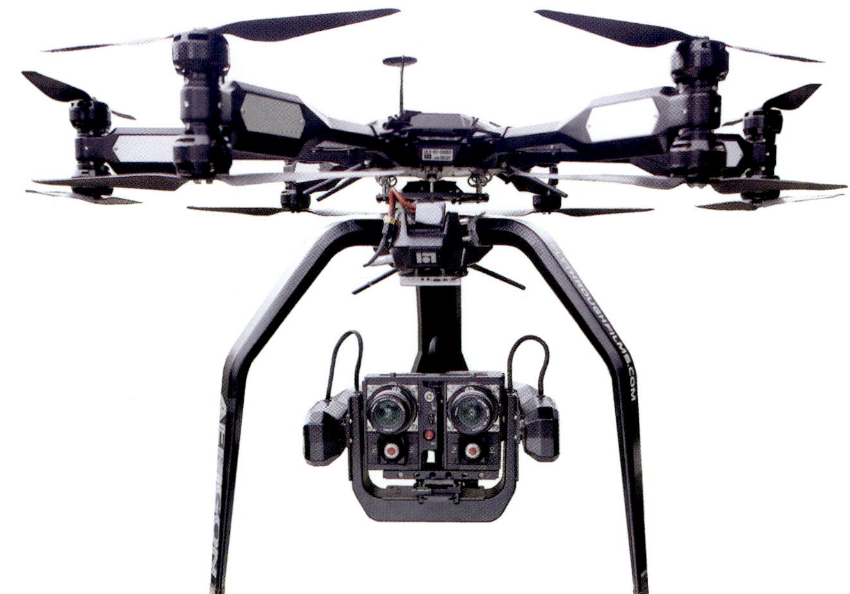

◀ Twin Red Epic cameras mounted on an Aerigon drone. This setup is used for shooting cinema-quality stereoscopic 3D video, although any crash would be an extremely expensive mistake!

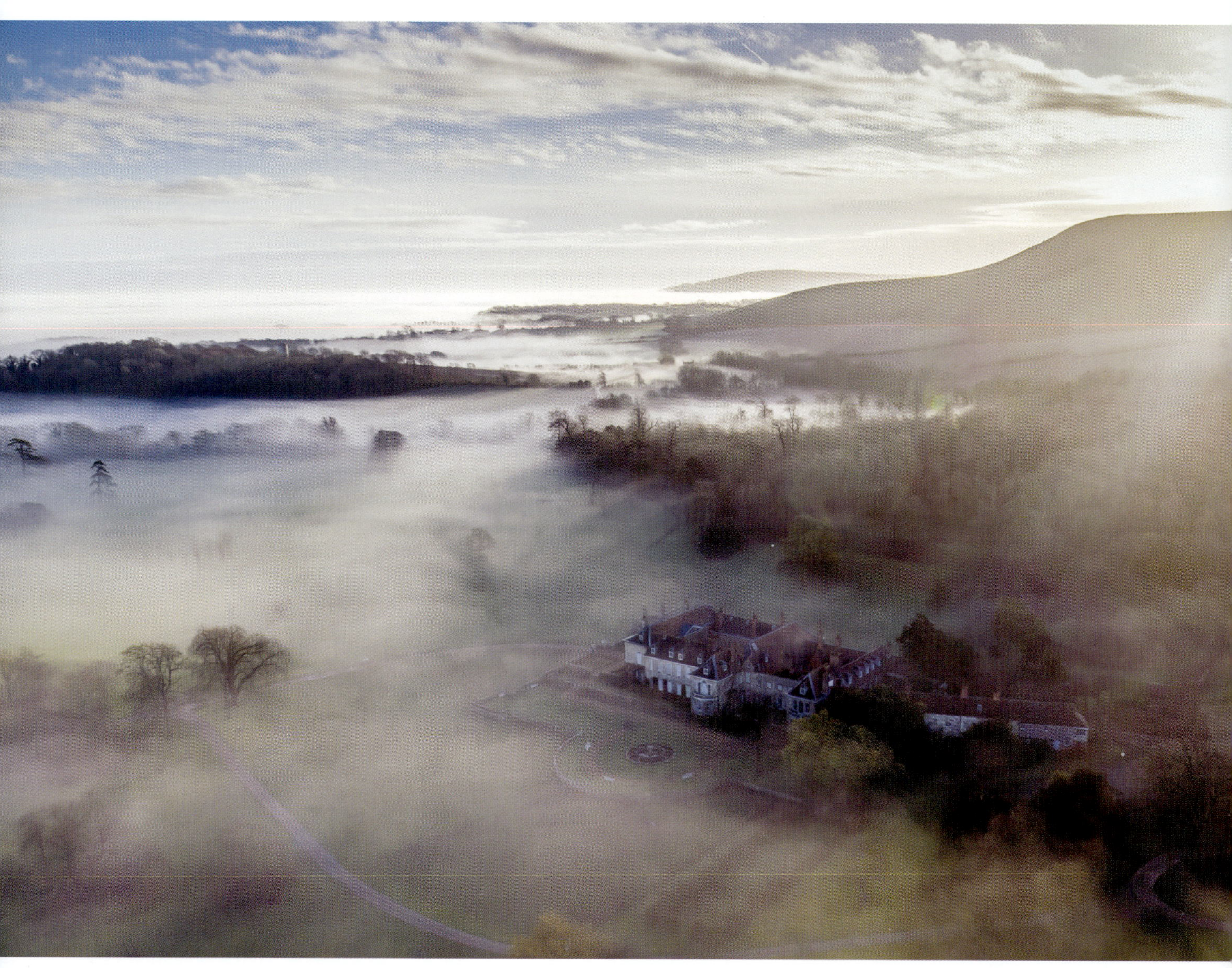

▲ A high-quality camera will allow you to capture beautiful scenes like this with maximum detail and dynamic range. This foggy shot was taken using a DJI Inspire Raw fitted with an Olympus 12mm f/2 lens.

MONITORING

You can have the best camera in the world on your drone, but it will be close to useless if you can't see what it's seeing in order to compose your photographs and videos. On a modern integrated camera drone this is usually achieved by sending a live video feed via the 2.4 GHz control frequency. This video display—along with vital flight data—can then be displayed on a smartphone, tablet, or an integrated display in the controller. This system is generally pretty easy to set up and has the added advantage that you can change camera settings via the touchscreen on the device.

On high-end setups, where conventional cameras are used, the video feed is usually either a 2.4 GHz digital link or an analog feed over 5.8 GHz. These are often displayed on a dedicated video monitor via HDMI or analog video inputs. Depending on your exact setup you may or may not be able to see camera data and flight data on your display as well.

It is worth noting that with both systems, the range of the video feed may be much lower than advertised and will often be lower than the control frequency link. Once the drone is a certain distance from the video receiver you may find the signal starts to break up or degrade. The proximity of obstacles and Wi-Fi or other radio frequency interference may also play a part here.

Another factor to look out for is the exact line of sight to the antenna. Check whether the receiver antenna are facing the drone, and whether the transmitter antenna on the drone are likely to be obscured by other parts of the drone itself. Small changes in orientation of the receiver or transmitter antenna can often make a big difference to the video reception. I have experienced video links that will degrade and drop out over distances as short as 100 yards/meters with a clear line of sight, but others that still display a clear HD video picture at distances of several miles or kilometers.

▼ One of the most common ways of seeing the drone camera's view, changing settings, and reviewing flight information is via an app on a tablet or phone connected to the remote.

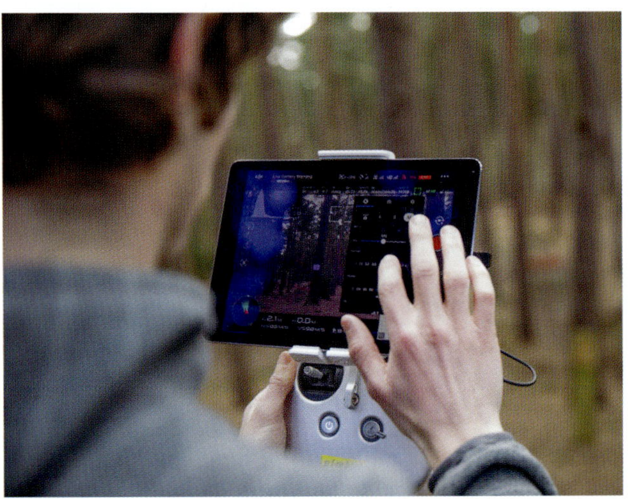

▼ A good monitoring system allows the camera operator to see the camera view as close to real-time as possible. A sunshade will help reduce reflections.

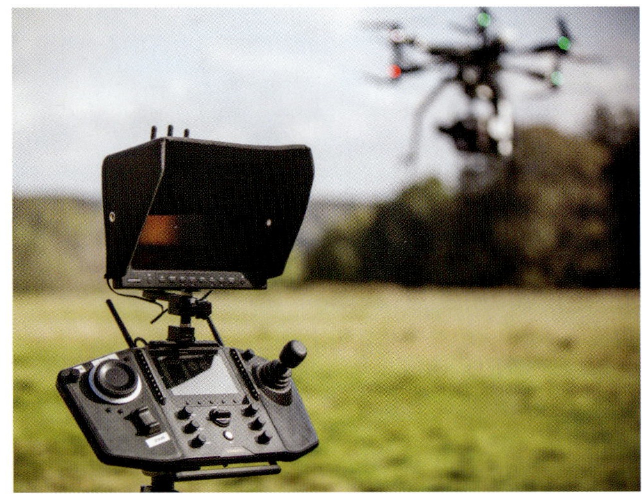

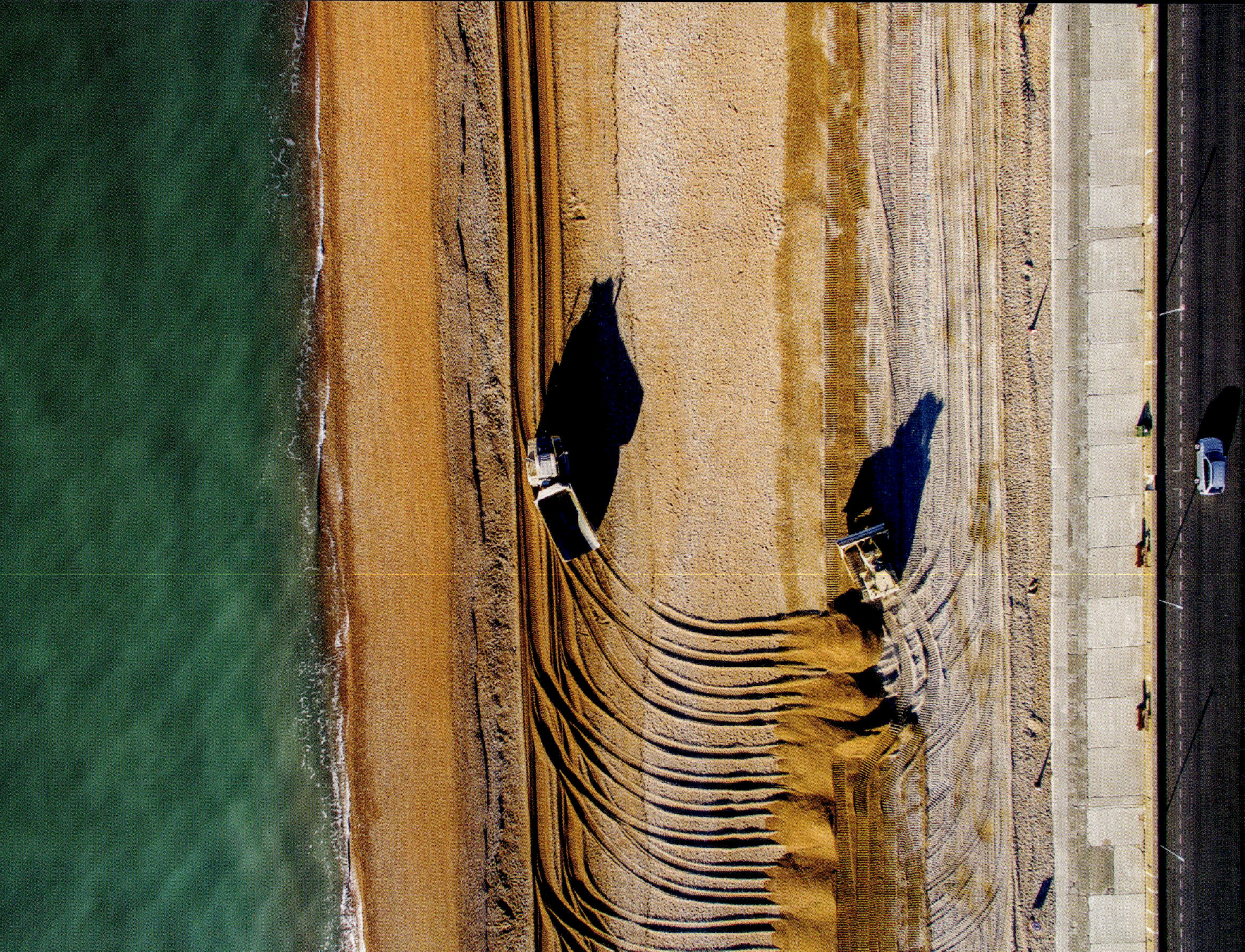

▲ In good conditions, without obstacles, even an analog video feed should give you a clear picture up to at least 500 yards/meters. The latest digital systems can deliver a high-definition feed over a much greater distance, although physical obstacles and radio frequency interference will curtail both the range and picture quality.

DUAL CONTROL

Some dual-operator drone systems come with an additional camera for the pilot. If the camera operator is looking in one direction with the main camera, but the drone is moving in another direction, the additional forward-facing camera can be used by the pilot to get a better view of where the drone is heading.

BATTERIES

A small- to medium-sized drone with an integrated camera will typically need little maintenance: it's a sealed unit and the end user has no access to its inner workings. However, one crucial component that always needs some attention is the battery.

Most drones run on Lithium Polymer (LiPo) batteries, which require a reasonable amount of care. LiPo batteries can become chemically unstable and even burst into flames if not treated correctly, so it's crucial to follow the manufacturer's guidelines on care and maintenance. Fortunately, many modern drones use either smart batteries or smart chargers that take care of most of the demands, ensuring all the cells in the battery are balanced. However, there are a few things you can do to get the best from your LiPo batteries and maximize their lifespan:

Charge safely: It's a good idea to charge batteries in a fireproof container and not to leave them unattended while charging. The biggest risk of battery damage and fire is during the charging process.

Keep track: LiPo batteries have a limited life span, which is generally measured in charge cycles. Some modern smart LiPo batteries keep a record of the number of charge cycles and the health of the battery, which you can check via an app. If these features are not available it's a good idea to keep a record of all your batteries, their charge cycles, and their performance (it will help if you label each one with a unique name).

Calibrating batteries: From time to time (perhaps every 20 charge cycles) you should run a deep discharge on the battery, which means running it down to 3.35V, or until it switches itself off. You can do this by leaving the drone switched on on the ground with its propellers removed.

Calibrate the battery by fully charging it and then discharging to 3.35V again. This will help the drone to report the percentage charge and voltage accurately.

▼ A typical charging setup, where one charger can supply the flight battery, the remote, and a phone or tablet.

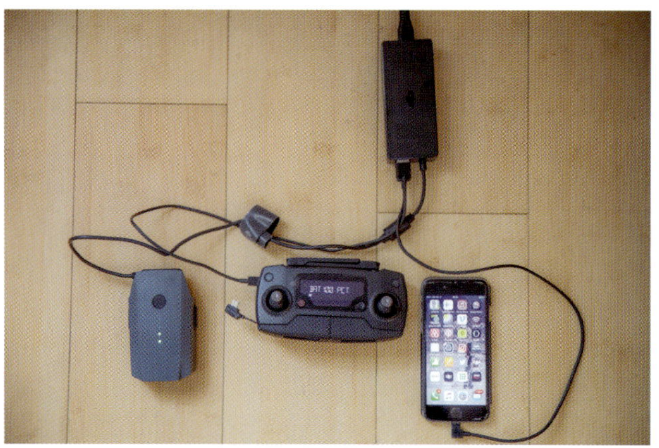

▼ It's a good idea to put the flight battery inside a LiPo-safe (fireproof) bag while charging. In the unlikely event of it catching fire, it could save you a lot of damage.

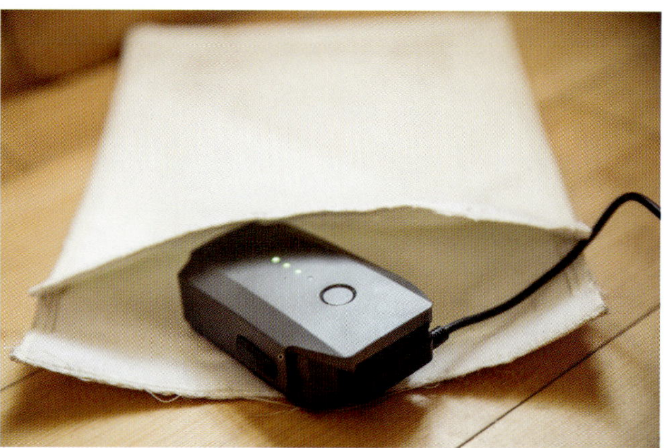

Keep them cozy: Before flying, your batteries should be kept warm (over 60°F/15°C). They will not work as they should below this temperature, so keep them in a warm pocket, or even invest in a heated battery case for very cold days.

Storage charge: You should not leave your batteries fully charged for extended periods of time. Ideally, fly within 24 hours of charging the battery. If you do not plan to use the batteries for an extended period, they should be left with a "storage charge" of 50–60%, preferably at room temperature (avoid extremes of heat or cold during storage). Many smart drone batteries will start to self-discharge after a number of days, but it is wise to check this has happened before extended periods without use.

Smart charging: If your drone batteries are not of the "smart" variety, invest in a high-quality smart charger, which will help extend the life of your precious batteries. These chargers have both discharge and storage charge features built in.

Fly smooth: Extreme maneuvers during flight will put a large strain on the batteries. If you go from full speed to a full stop very quickly, for example, this places a sudden demand on the batteries and it is when they are most likely to fail. Instead, try to accelerate and decelerate smoothly.

OTHER COMPONENTS

With a self-assembled drone a number of things should be checked on a regular basis:

- Check bolts and fixings are tight.
- Check wires and connectors for wear and looseness.
- Check propellers are tight and free from damage.
- Check motor bearings.
- Motors should be replaced or serviced after a certain number of hours (as recommended by the manufacturer).

▼ LiPo batteries do not work well at low temperatures, so should be kept warm until you use them. Keeping the battery in a warm car or in an inside pocket is a simple, low-cost option.

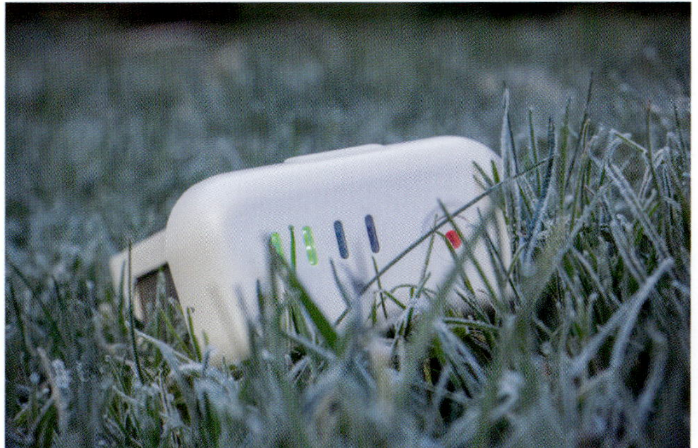

▼ When flying in temperatures below 60°F (15°C), you may want to use a battery heater to keep your LiPo battery warm.

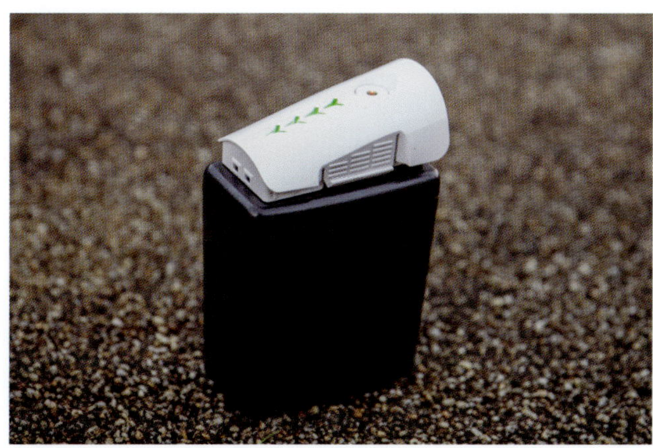

PROFILE: Eddie Oosthuizen

Eddie Oosthuizen is a software engineer and aerial photographer from Cape Town, South Africa, who started flying drones as a way to deal with an intense fear of heights. While growing up, he always avoided anything that involved not having his feet firmly on the ground, which meant he was denied the dizzying thrill of the aerial perspective. Now he uses a DJI Phantom 3 Advanced fitted with various ND filters to create abstract images of the world beneath our feet.

Eddie explains: "After getting my first drone, I spent most of my time just in awe of how different and fresh familiar places seemed. While looking onto the world from above, I found secret shapes in familiar places and beauty in the mundane. This top-down perspective levels out a busy world and turns our everyday 3D view into a 2D abstract masterpiece, which, if we were just strolling past on the ground, we would have never known existed. Through my drone photography I can show people a world they know, but in a way they might never otherwise experience."

▼ Two contrasting ground patterns collide on a border road just outside Pretoria, South Africa.

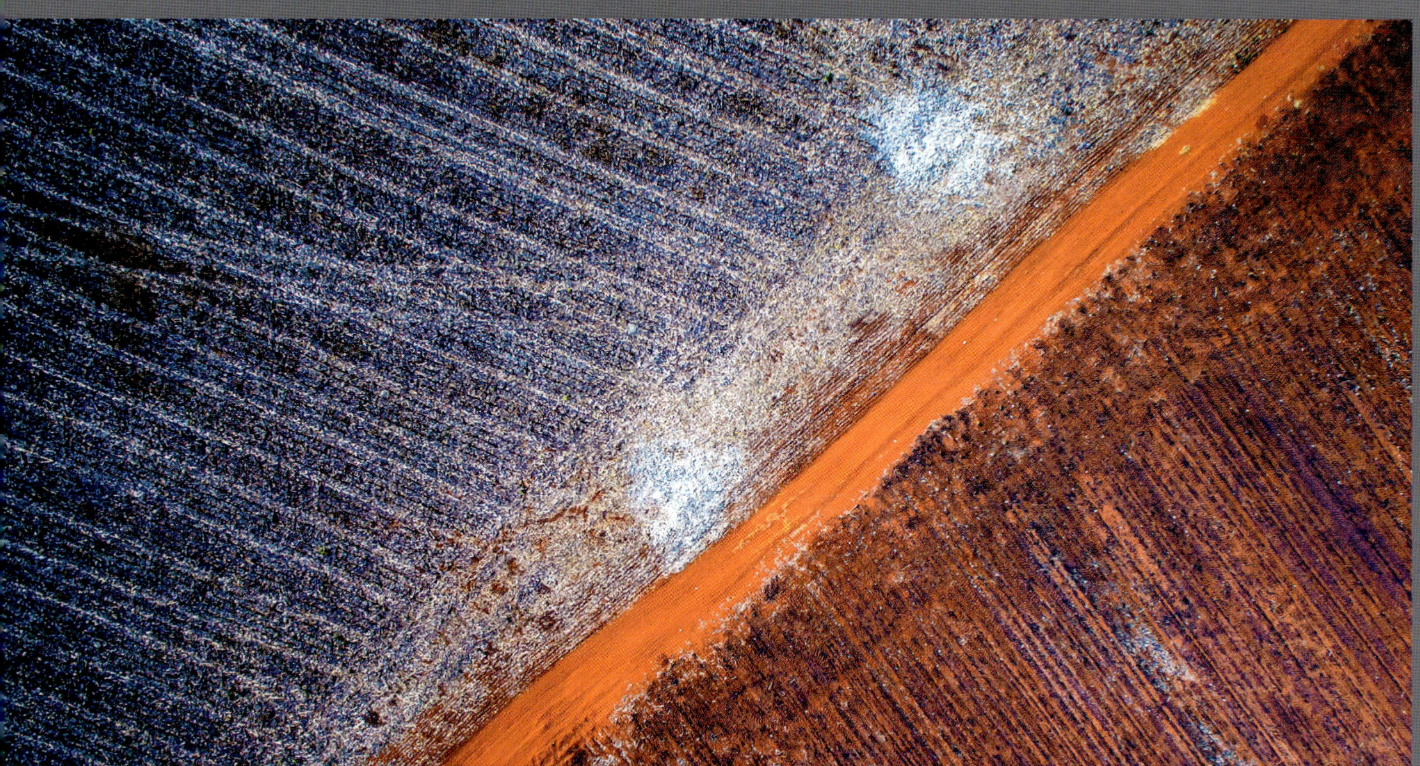

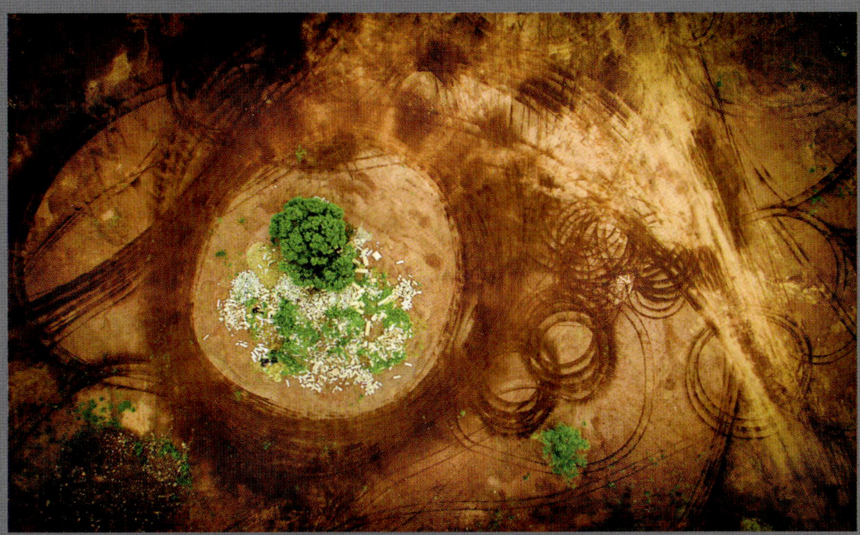

◀ The last remaining tree in a construction yard forms a small oasis in the dirt.

▼ A wave breaking on a rocky beach in Kommetjie, Cape Town, South Africa. An aerial viewpoint reveals a huge amount of color variation in the rocks.

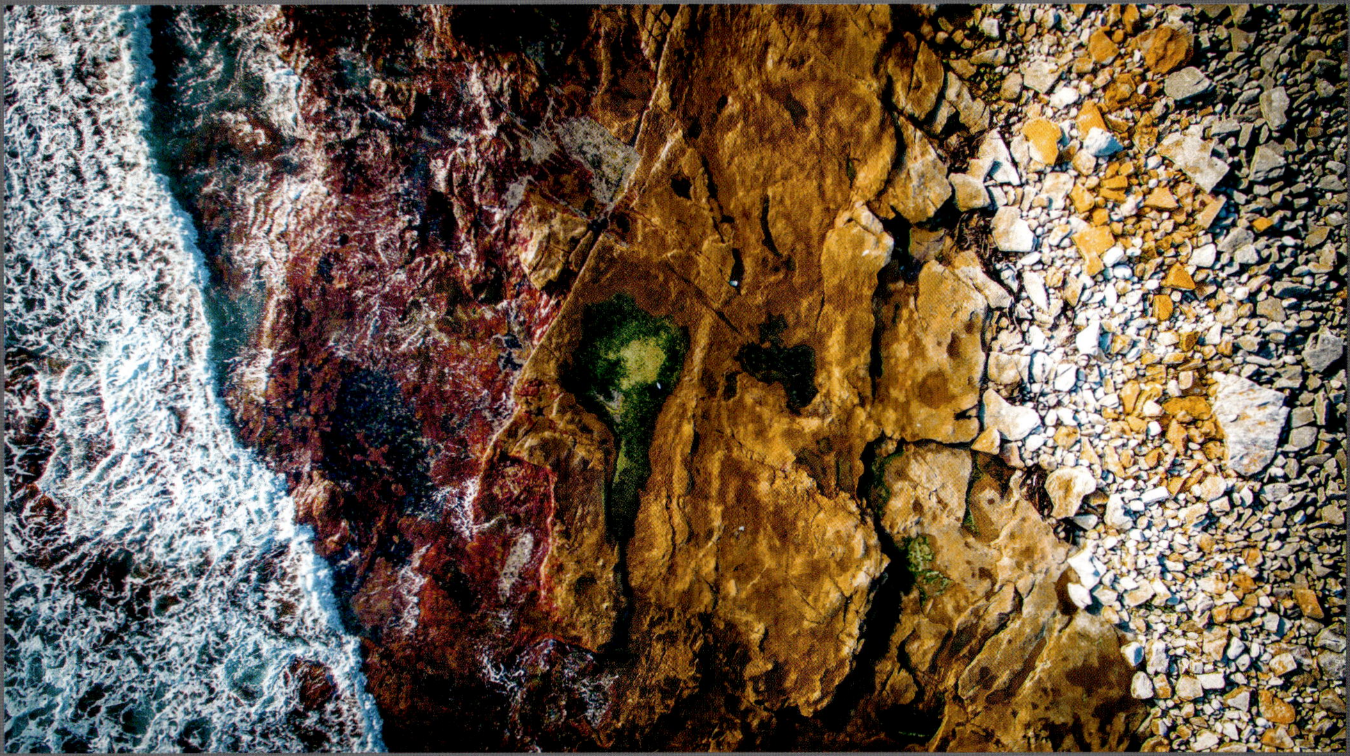

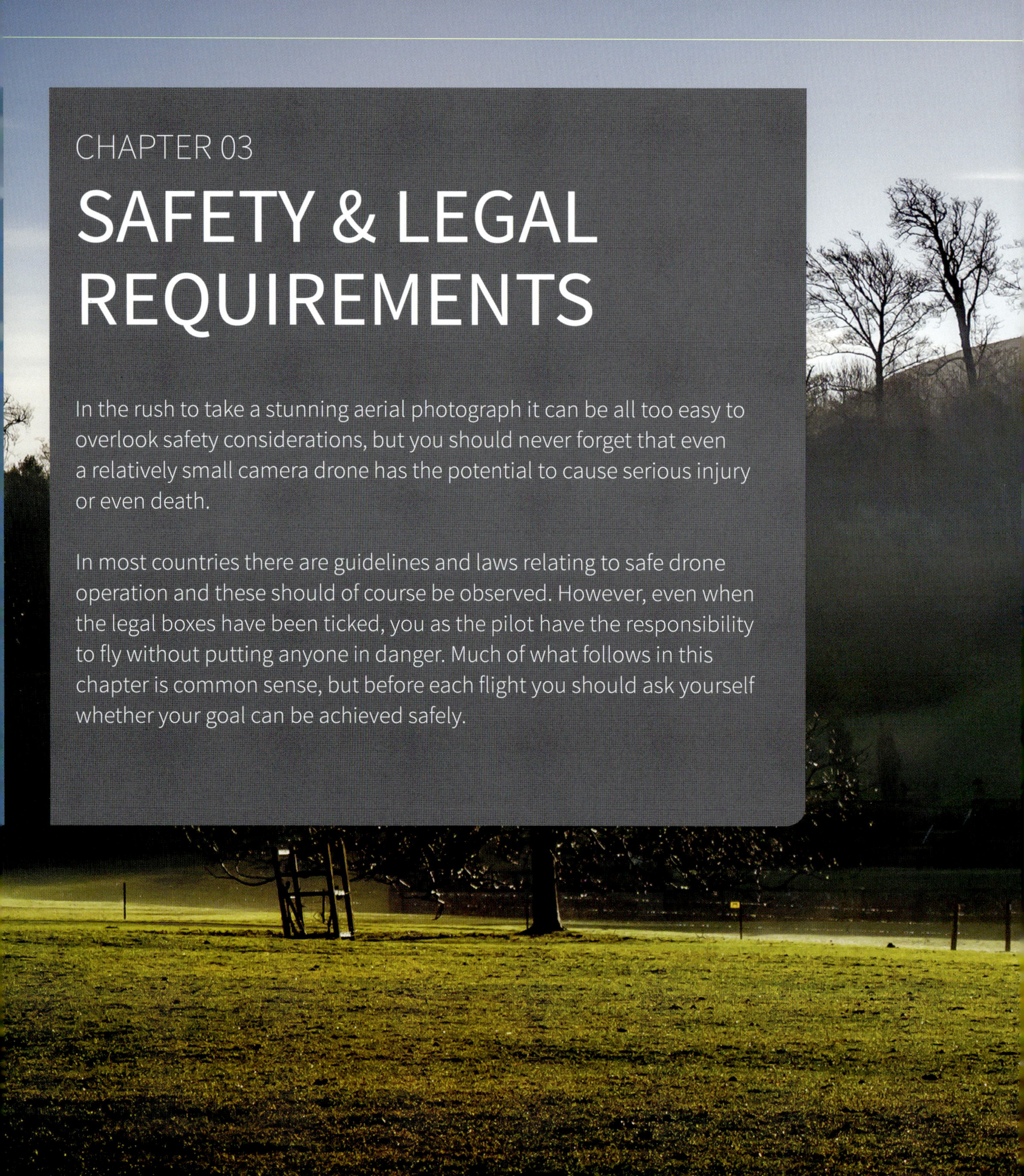

CHAPTER 03

SAFETY & LEGAL REQUIREMENTS

In the rush to take a stunning aerial photograph it can be all too easy to overlook safety considerations, but you should never forget that even a relatively small camera drone has the potential to cause serious injury or even death.

In most countries there are guidelines and laws relating to safe drone operation and these should of course be observed. However, even when the legal boxes have been ticked, you as the pilot have the responsibility to fly without putting anyone in danger. Much of what follows in this chapter is common sense, but before each flight you should ask yourself whether your goal can be achieved safely.

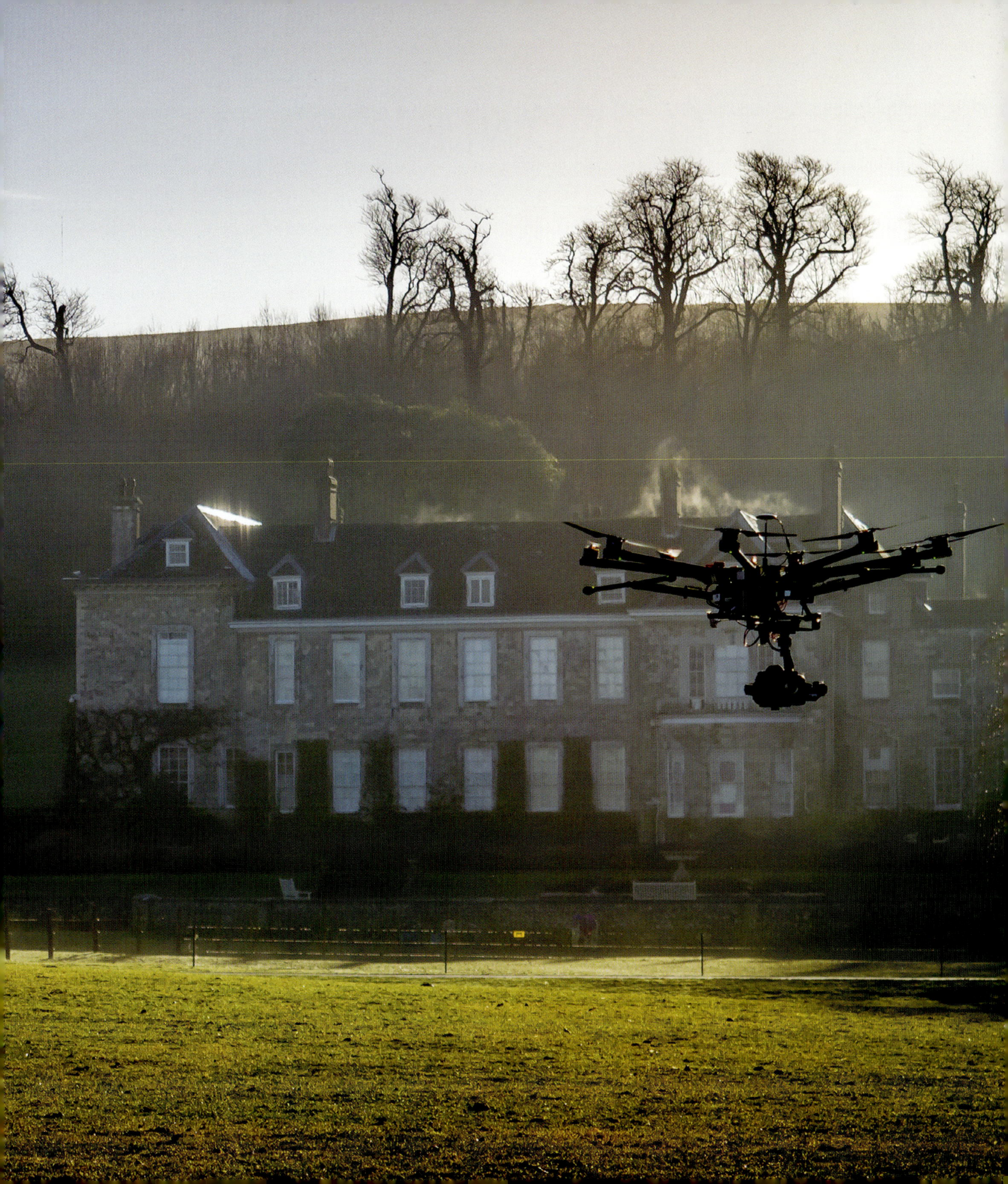

ESSENTIAL KNOWLEDGE

"This is not an unmanned aerial vehicle; it is a flying chainsaw." So begins the typical safety briefing by a drone pilot to a horrified location TV crew. It may sound a touch melodramatic, but a quick Google search for "drone injuries" confirms that even a relatively small drone has the potential to cause serious injury—even death—if flown carelessly. For this reason it is imperative that all drone pilots take their responsibilities seriously, acknowledge and minimize the risks involved, and are aware of and abide by the regulations in their particular country.

It is also important that drone pilots are sensitive to privacy issues. Drones have had some bad press, and although much of it is sensationalist and overblown, there are legitimate concerns over privacy.

SAFETY FIRST!

It may sound like a cliché, but the oft-repeated mantra of "safety first" rings particularly true for drone operations. Camera drones are small and very quick to set up and get airborne. There is often a tremendous temptation to see the opportunity for an incredible aerial shot and rush into trying to get it without first thinking about the safety (or legality) of the flight.

Before you get airborne, you should have a thorough checklist of both technical factors relating to the drone and environmental factors such as buildings, infrastructure, obstacles, airspace, and weather. You should also have a good understanding of the laws relating to drone operation in your country. Make a detailed and objective analysis of the risks; decide whether they are acceptable or not and what you can do to minimize or mitigate them.

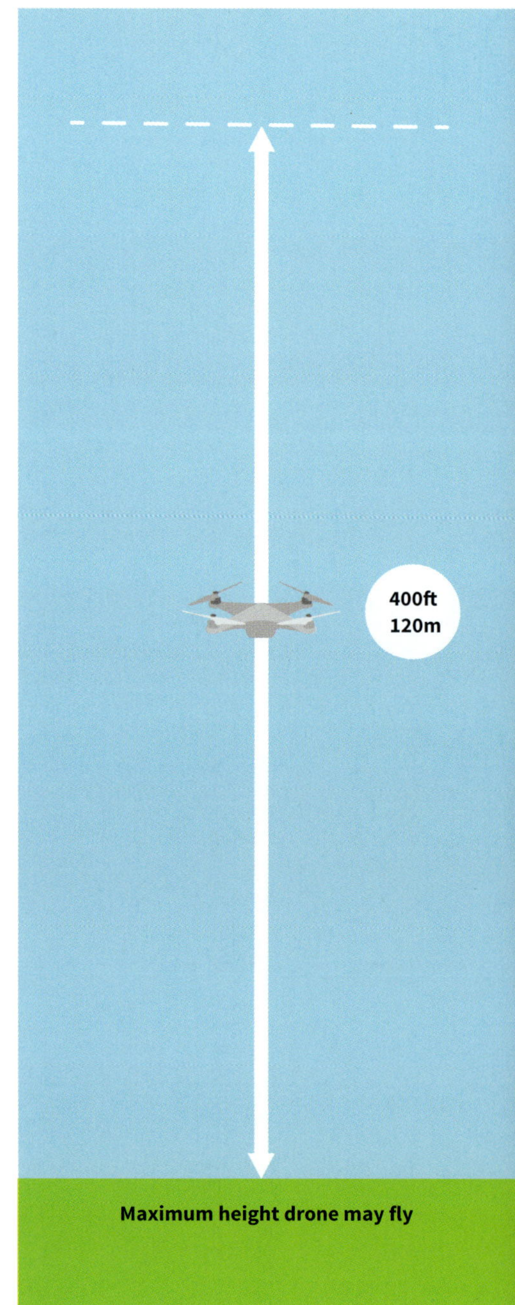

▶ Suggested exclusion limits for drone flying. The exact regulations vary from country to country, so you should always check the rules for the region you are flying in. However, keeping a safe distance from people and other obstacles—especially other aircraft—should be common sense.

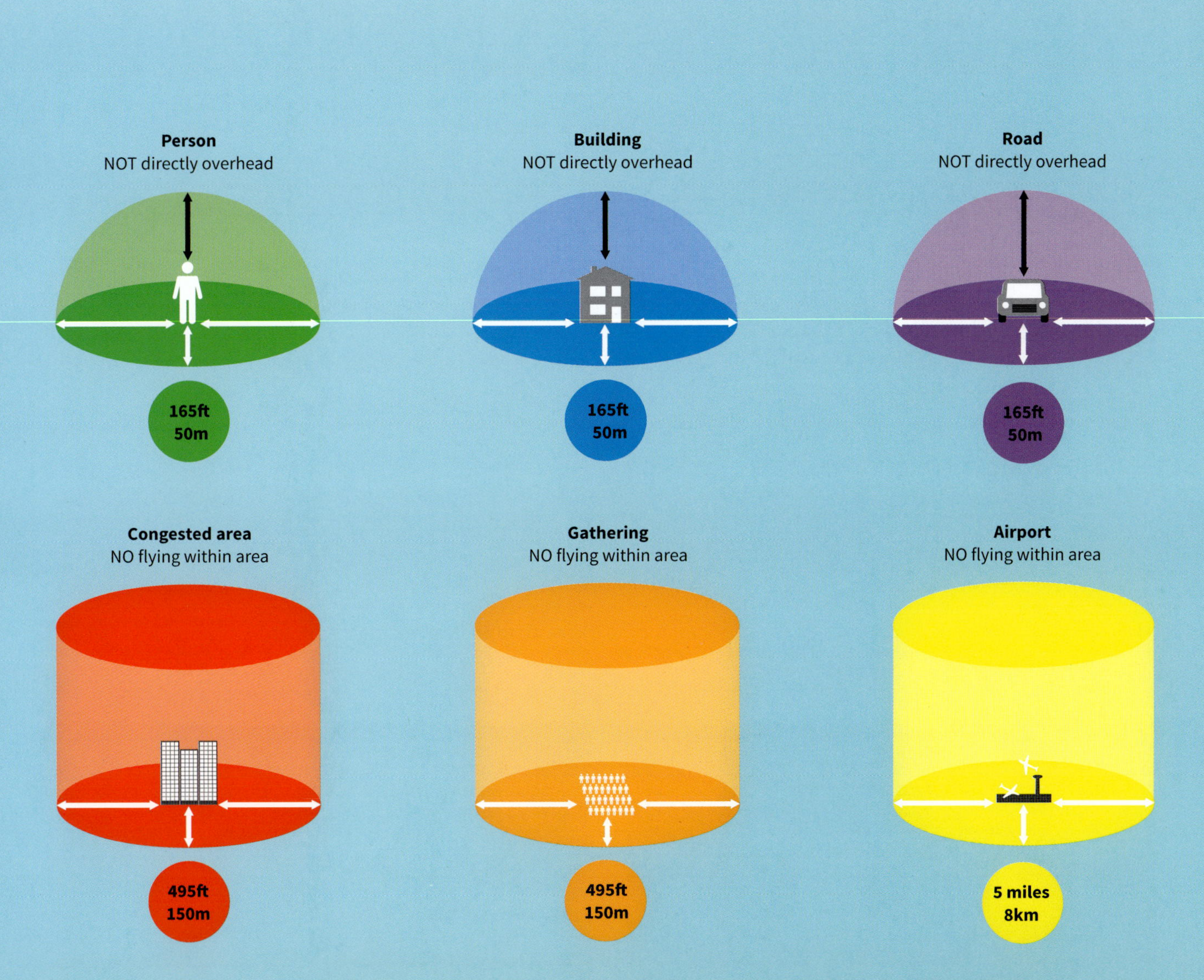
Person
NOT directly overhead
165ft
50m
Building
NOT directly overhead
165ft
50m
Road
NOT directly overhead
165ft
50m
Congested area
NO flying within area
495ft
150m
Gathering
NO flying within area
495ft
150m
Airport
NO flying within area
5 miles
8km

KNOW THE RULES

If you purchased your drone recently, there is a good chance it came with a leaflet in the packaging that outlined the regulations in your country. If not, there are links at the end of this book to a number of national aviation authorities' guidelines on drone use. Make sure you read them before your first flight. Failure to do so may not only endanger life, but also endanger the future of recreational drone use: national aviation authorities have been known to apply blanket bans on drone use on the basis of a single incident.

It is likely that the drone regulations will contain sections on airworthiness of your aircraft, the distances you need to be from the general public, buildings, and infrastructure such as roads and railways, as well as regulations concerning airspace to ensure adequate separation from full-scale aircraft and airports.

In some countries it is mandatory to register yourself and/or your drone with the relevant national aviation authority, regardless of whether you are a commercial operator or a recreational pilot. Again, it is your responsibility to know the rules and act accordingly.

COMMERCIAL OPERATORS

Pilots wishing to make money from flying their drone will often be subject to tighter regulation than enthusiasts. Again, this will vary from country to country, but many governments require aspiring commercial drone operators to register and obtain permission to carry out this type of aerial work. As part of the perquisites for obtaining this permission it is often necessary to gain a qualification, which may involve a written exam, a flight test, and development of an operations manual.

Typically, the written exam will cover topics such as regulations, aircraft maintenance, meteorology, safety, and insurance. The flight test is designed to ensure that you can plan and safely and confidently operate a drone, and deal with various malfunction scenarios and other unexpected events. The third element—the operations manual—is a comprehensive document detailing the specifications of your aircraft, pilot qualifications, relevant regulations, safety procedures, and your insurance details.

▼ Students studying theory at a Ground School for prospective commercial drone pilots. In most countries, such courses are a prerequisite to obtaining a permission for commercial operations.

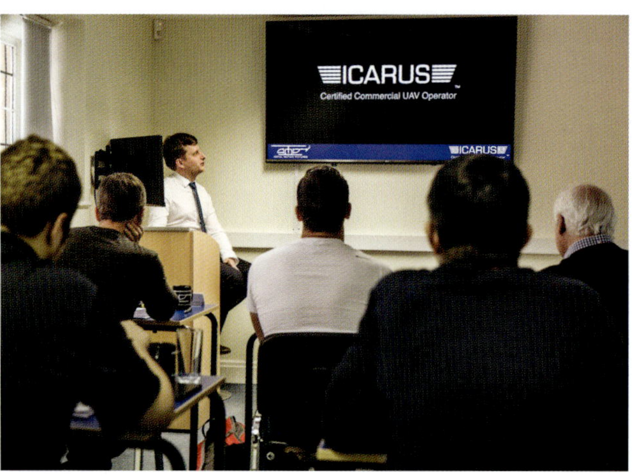

▼ A practical flight test underway. Students must demonstrate that they can operate a drone competently and safely.

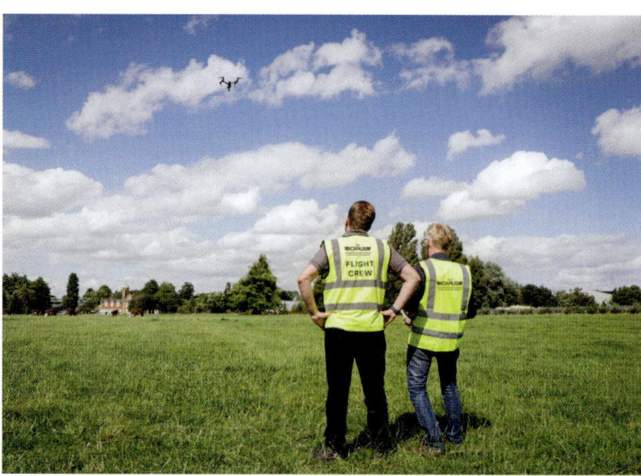

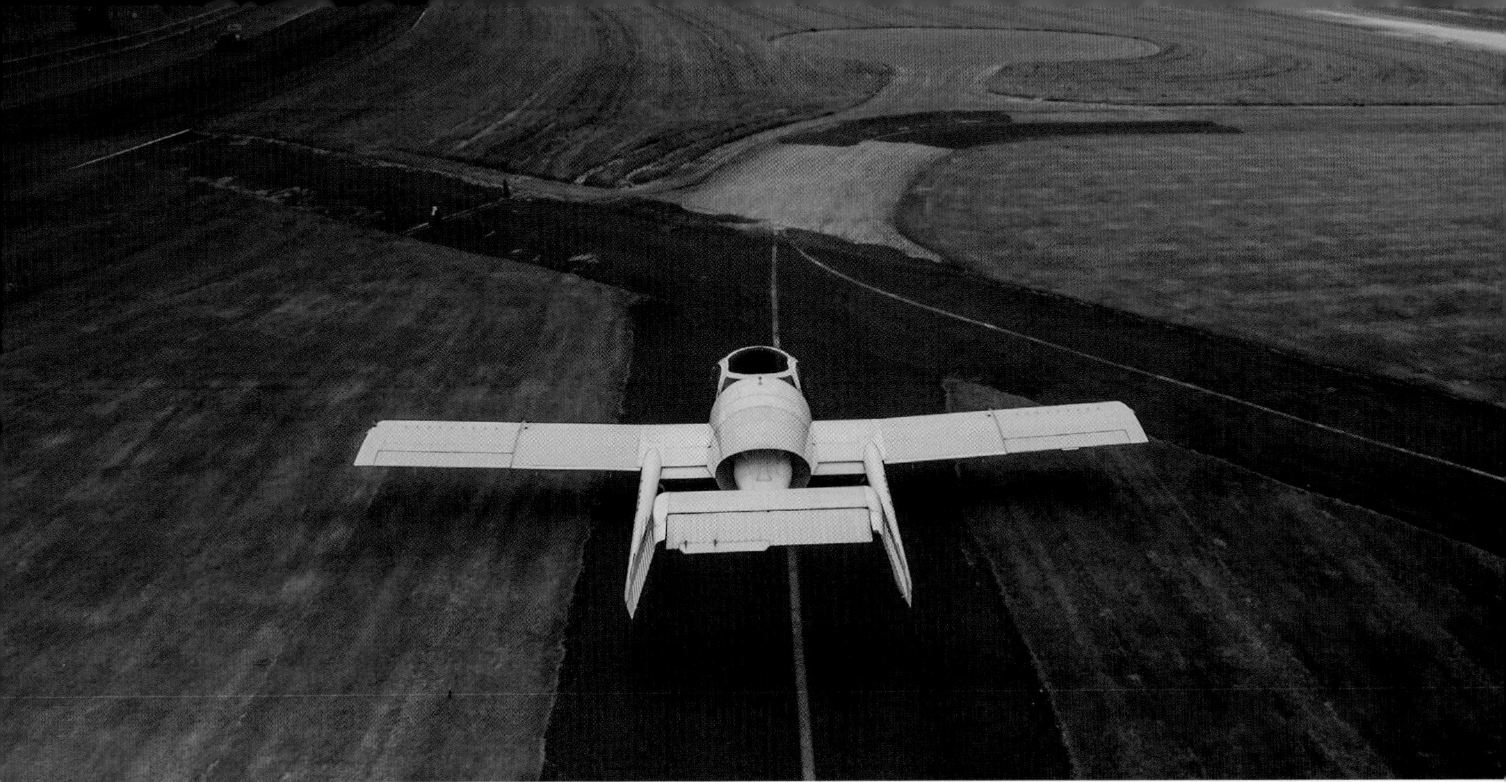

▲ I have been asked to fly drones on or close to operational airfields on several occasions. Each time, a prior conversation with air traffic control—followed by detailed communication on the day—has resulted in safe operations.

AIR TRAFFIC

Possibly the most serious safety concern involving drones is the chance of a collision with a manned aircraft. Many owners of recreational drones are not fully aware that they are governed by the rules of air navigation of the country in which they are operating. It's therefore essential that all prospective drone pilots familiarize themselves with the basic concepts of air traffic and how they apply to the use of camera drones.

All large airports have no-fly zones around them. Even with a small drone you should maintain a minimum distance of 5 miles (8km), but also double-check the regulations from the national aviation authority of your country. Many newer drones include "geo-fencing," which uses GPS positioning to effectively stop any drone from taking off within—or entering—any specified restricted areas.

As well as major international airports, there are also numerous smaller airfields and heliports to consider. Although these may not always appear operational, the simple rule is not to fly near them unless you have contacted the airfield in question and they have told you it is safe to do so.

Many countries will also specify a maximum altitude of 400ft (120m) for drones, and there is good reason for this. In most countries, light aircraft in transit will stay at least 500ft (150m) from the ground directly beneath them. However, drone pilots should always remain vigilant for manned aircraft, as helicopters, sailplanes, and paragliders are often encountered flying lower than 400ft.

The safest course of action if you hear or see an approaching aircraft is to get your drone down to ground level as fast as you safely can. Above all, if you have *any* doubts about possible interactions with manned aircraft you should contact air traffic control at the nearest airport or airfield.

KEEPING YOUR DISTANCE

How close you are comfortable to fly to people, buildings, and cities will vary according to your experience, the weather conditions, the size of your drone, and numerous other factors. There are also hard and fast regulations governing where and when you should fly in relation to your immediate environment.

People: Whatever the regulations are in your country, it is always a good idea to maintain a safe distance between your drone and members of the public: in the UK, the minimum distance is 165ft (50m).

It is not a good idea to take off if there are large numbers of people around and you think it may be difficult to return and land while still maintaining this safe clearance. You should also avoid flying directly over the heads of members of the public, even at altitudes greater than the minimum distance: if your drone fails, it is only going to go in one direction, and that's straight down. If there is a large crowd of people, such as would be found at an event or outdoor gathering, the separation distance will generally increase—500ft (150m) is the minimum distance in the UK, with no direct overflight.

Roads: A similar clearance to members of the public should be given to roads. As well as the risk of direct collision with road traffic, a drone could easily distract a driver, indirectly resulting in an accident. Where a drone needs to fly close to a road—a commercial pilot filming a car commercial, for example—permission should be sought to temporarily close the road for the duration of operations.

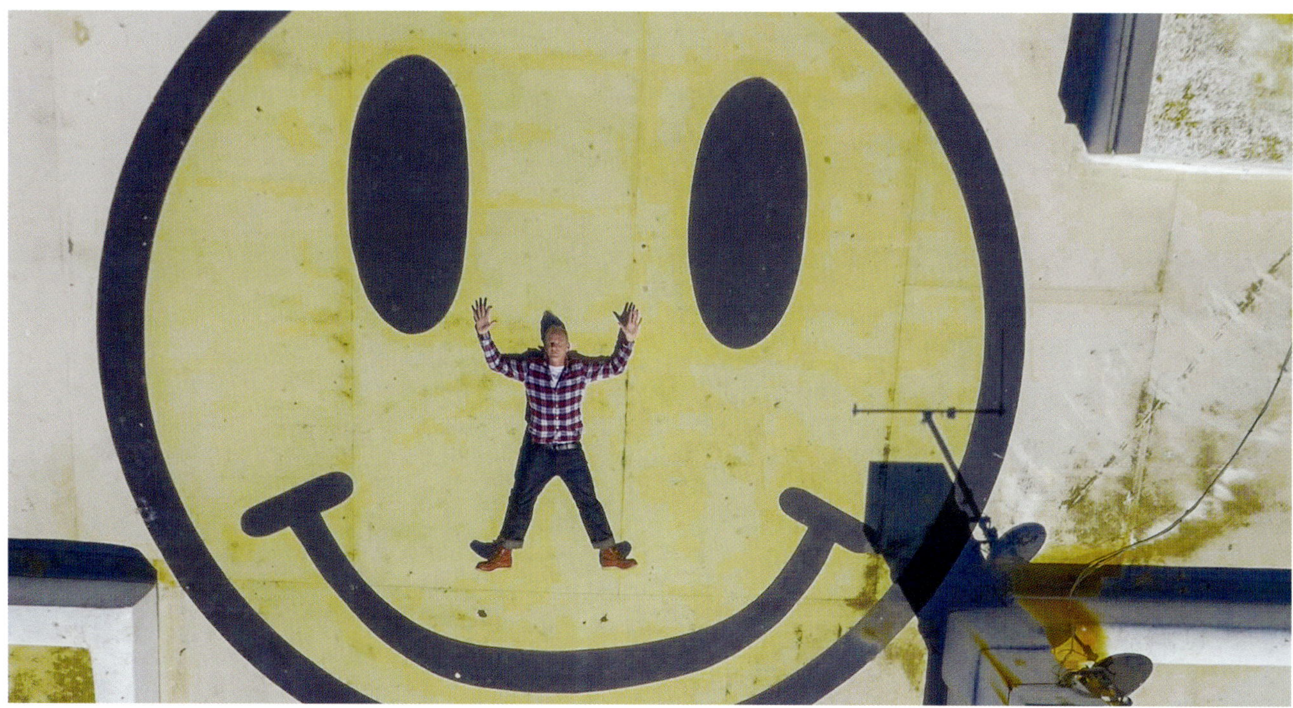

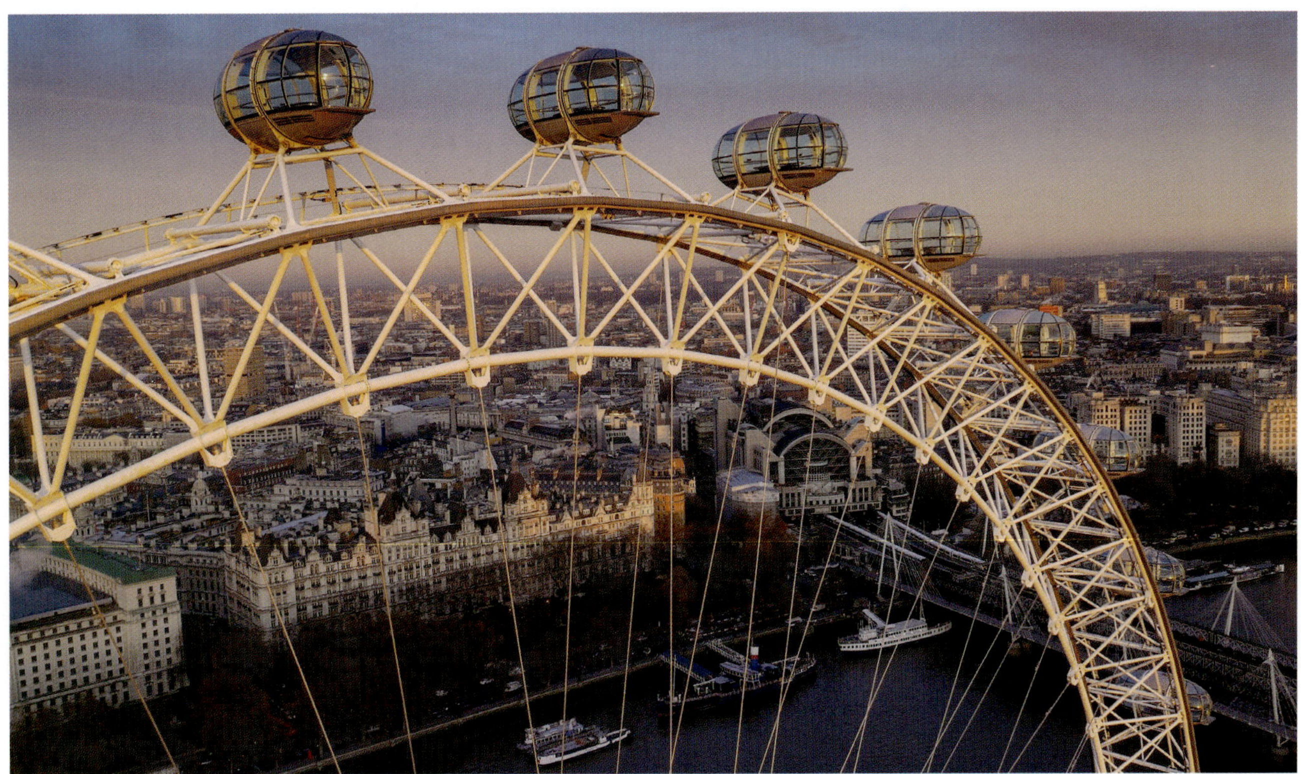

Buildings: For reasons of both privacy and safety, many countries enforce a regulation that drones should maintain a distance of at least 165ft (50m) from buildings not in their control. If you have control of a building and permission from the owner (including control of any people entering and exiting the building), it is permissible to fly closer than this.

▲ Flying in—or close to—congested areas is obviously high risk. This shot of the London Eye (from Fly Through Films) was only possible after extensive consultation with the police and local authorities.

◀ If a person is under your control you may operate a drone closer than 165ft (50m), but you should still try to minimize flying directly overhead. This is DJ Fatboy Slim in a music video shoot.

Congested areas: Congested areas are defined in relation to a city, town, or settlement as any area that is substantially used for residential, commercial, industrial, or recreational purposes. Be particularly wary of areas with a high density of human population, both traffic and/or pedestrians.

Drone pilots should avoid flying close to or directly over any congested areas, with suggested limits of 165ft (50m) for drones weighing up to 15lb (7kg), and 500ft (150m) for heavier aircraft. The reason for this is that any malfunction and resulting crash is far more likely to result in the drone coming into contact with a person or vehicle in the congested area.

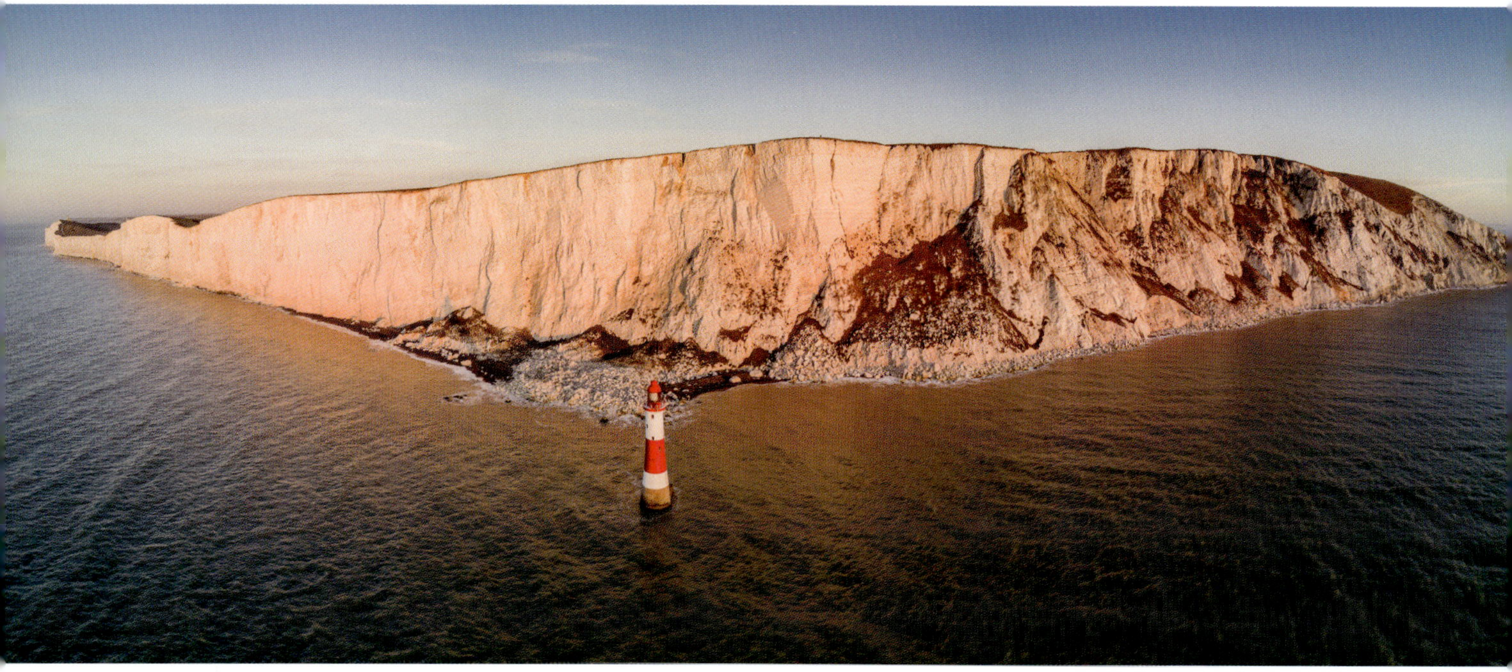

▲ Taking off from a boat (and landing back on it) can be a good way to get great shots, but you need a certain amount of planning and skill if you want to achieve it safely.

▼ Flying a drone over the sea comes with its own set of challenges and risks—mistakes can potentially result in you losing your drone.

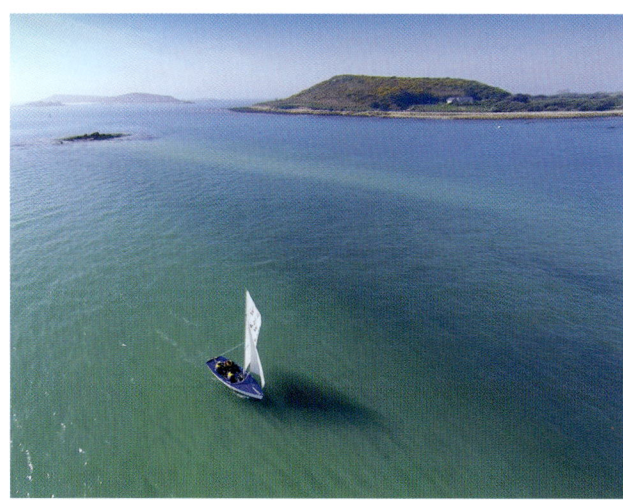

Sea & boats: When flying drones over water and near boats, there are some special considerations to bear in mind. Firstly, if you crash your drone in the water, you probably aren't going to get it back, and even if you do it probably won't be repairable.

Secondly, if you take off from a boat and the boat moves while the drone is in the air, the "return to home" location will be open water. In the event that you lose radio contact with your drone, it would try to "return home" and land in the sea. Fortunately, many newer drones offer a "dynamic home point," which means the drone constantly updates the "home" location via a GPS receiver in a connected phone. This is not infallible, but it does give you more of a chance of rescuing your precious drone in a signal-loss scenario.

Finally, before setting off to fly a drone from a boat, give some thought to how much space you need to take off and land. Even if you are happy to take off and land a small drone on a boat deck, if the sea is rough and the boat is moving, landing will be difficult. One option here is to "hand-catch" the drone, although this is really only for experts.

Wildlife & livestock: It's a good idea to be sensitive to any wildlife or livestock in the area you want to fly: a collision with a bird may not end well for either the bird or the drone. You should certainly keep clear of sensitive wildlife areas, and be aware that flying drones is often subject to a seasonal or permanent ban in nature reserves or national parks. While it seems many birds largely ignore drones or are mildly curious, breeding birds can be easily spooked or even become aggressive—drone pilots have been successfully prosecuted for harassing wildlife.

Similar rules apply to livestock and domestic animals: be sensitive and keep your distance. Horses, particularly, are spooked easily by new sights and sounds, and have been known to be severely injured or even die as a result of injuries sustained as they panic and bolt.

Dogs can also be unpredictable around drones, often chasing them and even jumping to try to catch them. I have had instances where a dog has been over 165ft (50m) away when I've been approaching for a landing, but it has reached the drone in seconds. At this point I've needed to quickly abort the landing and get the drone out of the dog's reach until the dog is back under its owner's control.

▼ You should maintain a safe distance from livestock or other animals and always be wary of causing any distress.

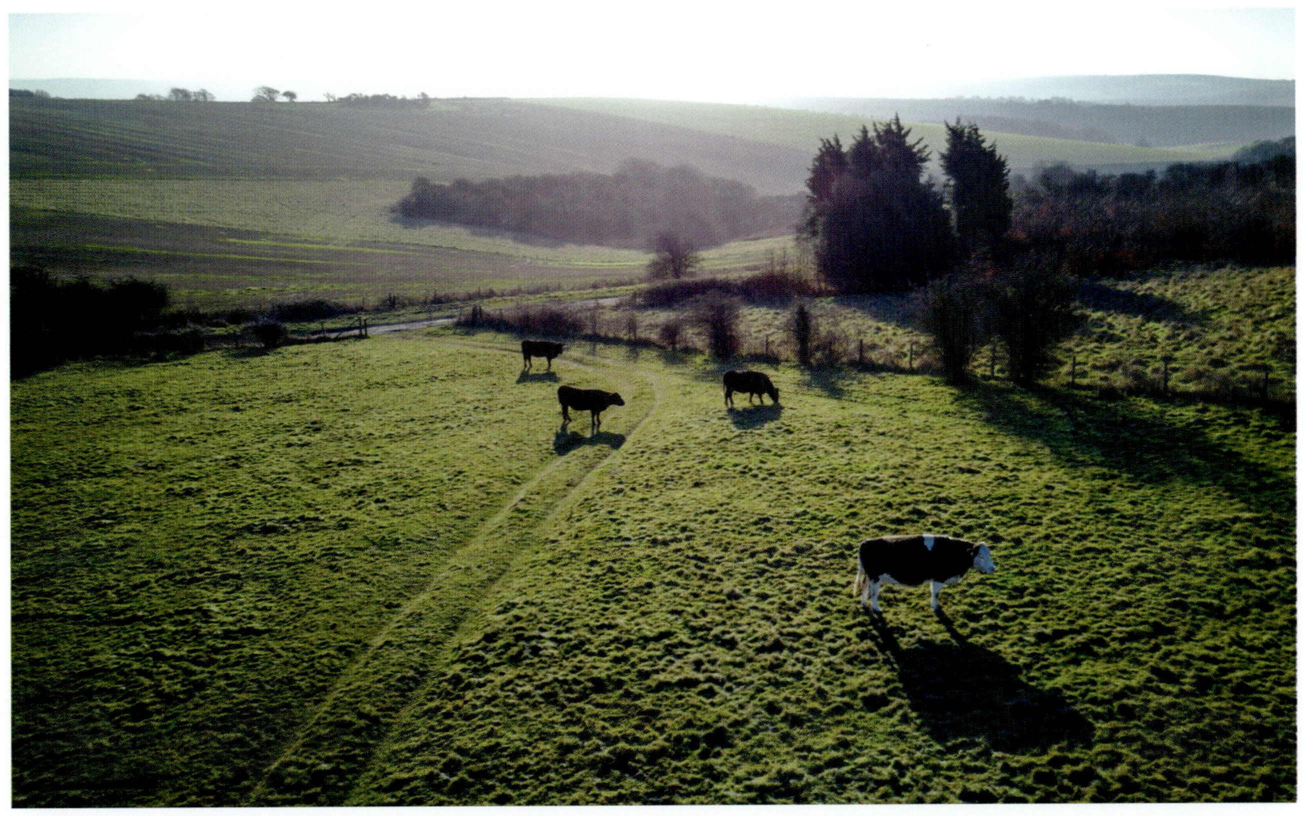

OTHER LEGAL REQUIREMENTS

Quite apart from sticking to the rules concerning proximity to people and objects, you also need to have a good awareness of stationary and moving objects around your drone.

LINE OF SIGHT

In many countries it is a legal requirement to have a direct line of sight to your drone. Although it is possible to fly a drone that is out of sight, relying on the drone's camera view via your display, this is generally not recommended. If you are flying sideways or backward relative to the camera direction, then a stationary obstacle will go unnoticed, while your perception of the proximity of obstacles will be distorted, particularly if you are using a wide-angle lens. Having a direct, unimpeded view of your drone will also allow you to identify other hazards that you might not notice through the camera—a flock of birds approaching from the side or behind, for example.

My own preference is to be as close as I practically can be to the drone for each shot. This becomes increasingly important the closer your drone is to obstacles. If I need to fly close to trees, buildings, or other obstacles I will move myself as close as possible, as the nearer you are, the better your depth perception is. There have been many drone accidents where a pilot is flying fast toward a building or tree and slammed straight into it. There is even a famous YouTube clip involving a painful-looking drone-human interface on a wedding day, where the drone pilot managed to hit both the bride and groom square in the head. Luckily the couple escaped without serious injury.

For pilots who really want or need to fly via first-person view alone, it is strongly recommended that they use an observer. This should be someone standing next to them (or at least in constant contact by radio) who *does* have direct line of sight to the drone and can alert them to any hazards.

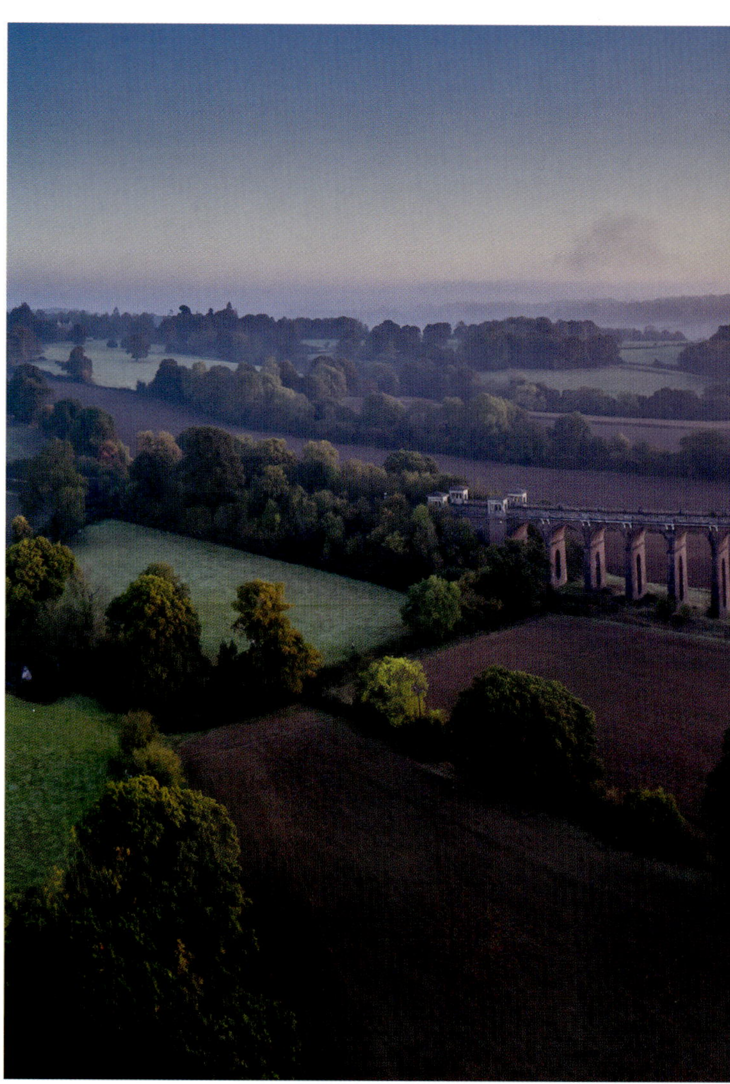

▲ Some obstacles are easy to see, such as trees, bridges, buildings, and other structures, but judging your *proximity* to them can be much harder to gauge. Having direct line of sight helps, and the closer you are to your drone, the better your depth perception will be, which enhances your ability to fly precisely.

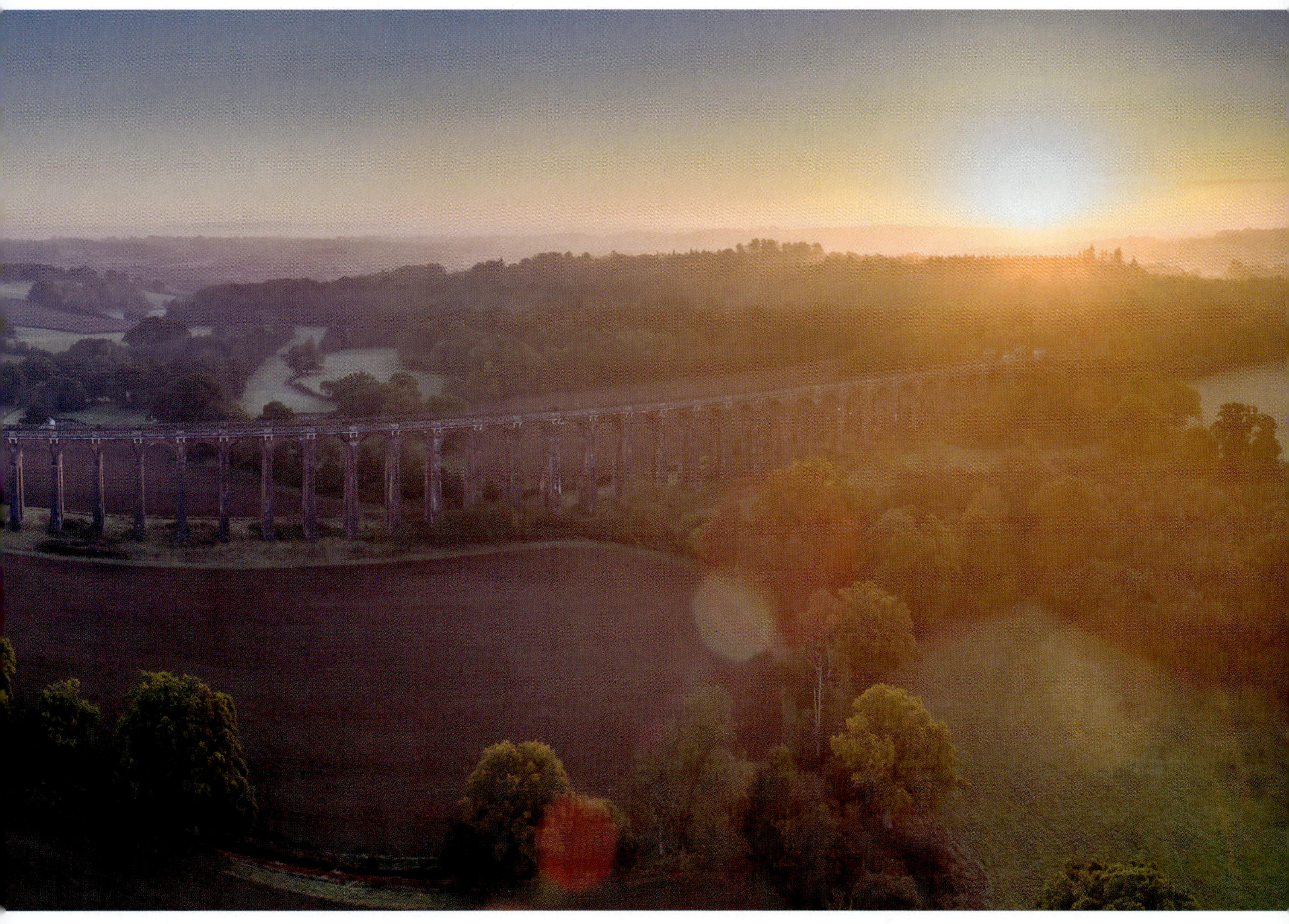

RETURN TO HOME

Return to home (or RTH) is a feature where the drone "remembers" its take-off location. In the event that you lose communication with the drone during a flight, the battery level gets critical, or you simply lose sight of the drone and want to get it back, RTH will automatically return your drone to its launch spot and land it.

For RTH to work, the drone needs to obtain an accurate GPS position while it's on the ground, prior to take off. This can take a few seconds and it's essential that you wait for confirmation that the take-off location has been recorded before you take to the air.

In many countries, this feature is a legal requirement for commercial drone operations and having a system where any failure in communication on the control frequency—or a low-battery warning—can trigger an RTH is a great feature. However, it can also have unexpected consequences, so before a flight you should always consider what might happen if an RTH is suddenly triggered.

There are plenty of examples of RTH mishaps online, including a clip where a drone pilot was flying on the far side of a huge rock stack, of the type found in Monument Valley, USA. With so much rock between the transmitter and the aircraft, the drone lost communication with the controller and initiated a return to home. This was not a modern drone with obstacle avoidance, though, so it attempted to take the most direct route back to its take-off point, barreling straight into the rock face!

Another example happened to a friend of mine (who shall remain anonymous!). He was stood on a riverbank, getting a great shot as he flew his drone between the concrete supports of a large bridge spanning the water. Unfortunately, when the drone was some way out—and directly under the bridge—it lost transmission. Even though a straight line return would have got it back safely, the RTH altitude was set 65ft (20m) above its take-off altitude, so the drone immediately ascended. In the process, it collided with the underside of the bridge before tumbling into the river below.

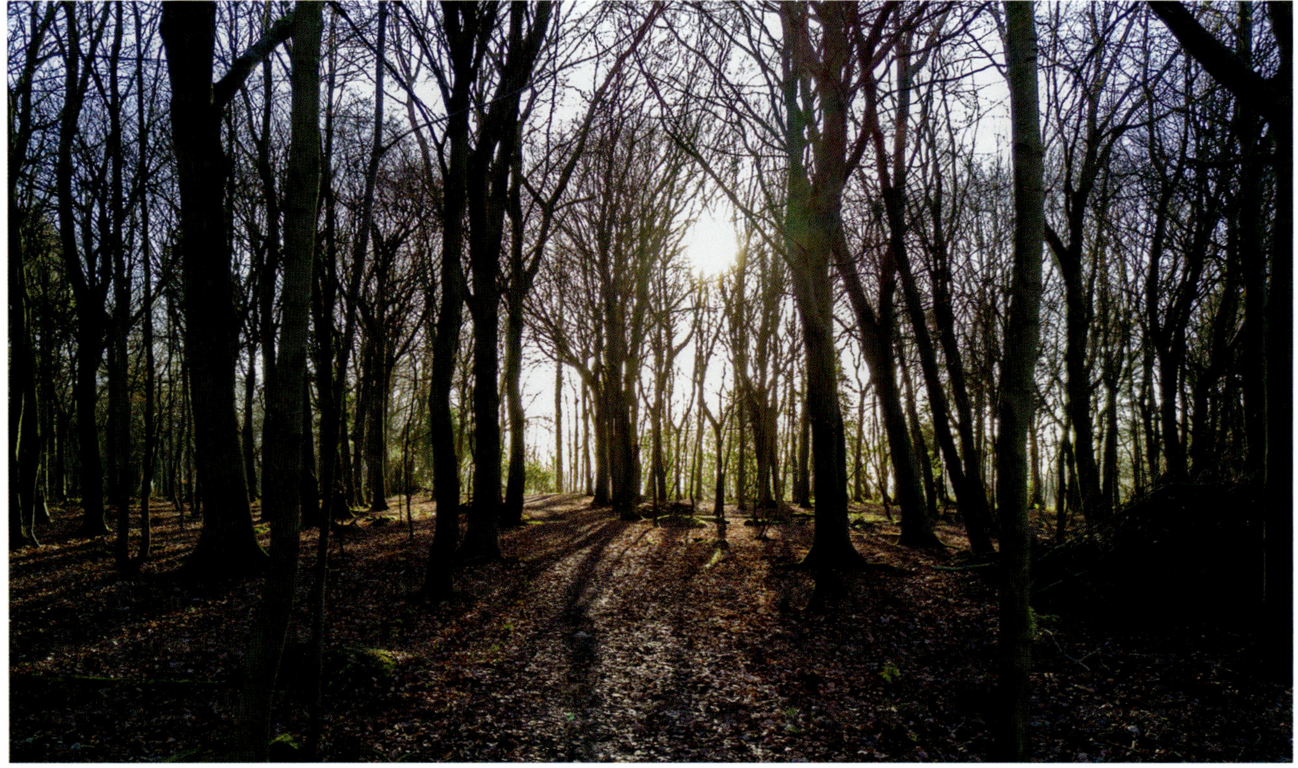

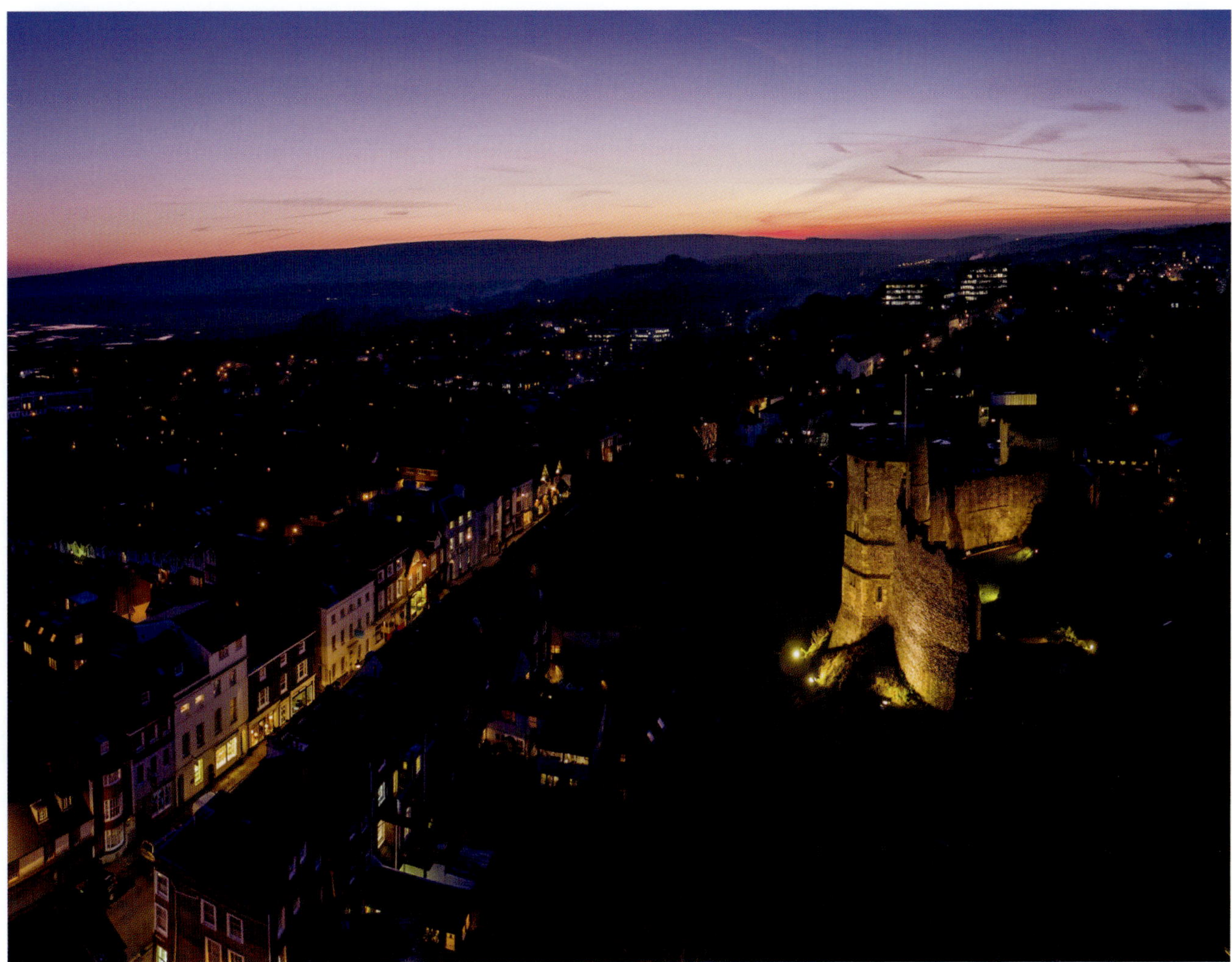

◀ Flying in woodland can be fun and challenging, but if you lose communication and the drone tries to return to its take-off point automatically, there is a fair chance it will collide with a tree!

▲ Flying at night comes with its own set of challenges. Check the regulations for night flying in your area, and always ensure you have a good line of sight and an illuminated take-off and landing area.

NIGHT FLIGHTS

There are obvious additional hazards involved when flying a drone at night, and in many countries it is simply prohibited or restricted. Where it is permissible, you need to have lights on your drone that will not only help you see it, but should also clearly indicate its orientation. It is also essential that you light the take-off and landing area. It will pay to make a familiarization visit during daylight hours, but you should still proceed with caution and know your local regulations.

PRIVACY

With the proliferation of surveillance and military drones, it is perhaps unsurprising that the public can treat these "eyes in the sky" with some degree of suspicion or even hostility. The vast majority of my encounters with members of the public have been positive—people are generally enthusiastic and interested—but you need to be aware that there may be people who are suspicious or even scared of drones. If you are approached by a member of the public who is not happy with what you are doing, land your drone at a safe distance and then explain the purpose of your flight. Try to remain calm and courteous at all times.

For example, I had been asked to photograph a house when a teacher from a nearby school marched up to me in a purposeful manner. The school playground was over 650ft (200m) away from the drone, but the children had spotted the drone in the sky and were excited. The teacher was rightly concerned about privacy issues, so I explained what I was doing and even showed her the video feed, which clearly demonstrated that the kids in the playground were barely visible, let alone recognizable. The conversation ended on good terms and she even invited me to give a talk on drone photography at the school.

The important thing to remember is that although you should always be a certain distance from any person or building not under your control for safety reasons, the public also has a reasonable right to privacy. If you are filming anyone—even on public land and from beyond the minimum "safe" distance—if they are recognizable you need to comply with any data protection laws in your country. This is no different to someone else using a camera or CCTV in a public place.

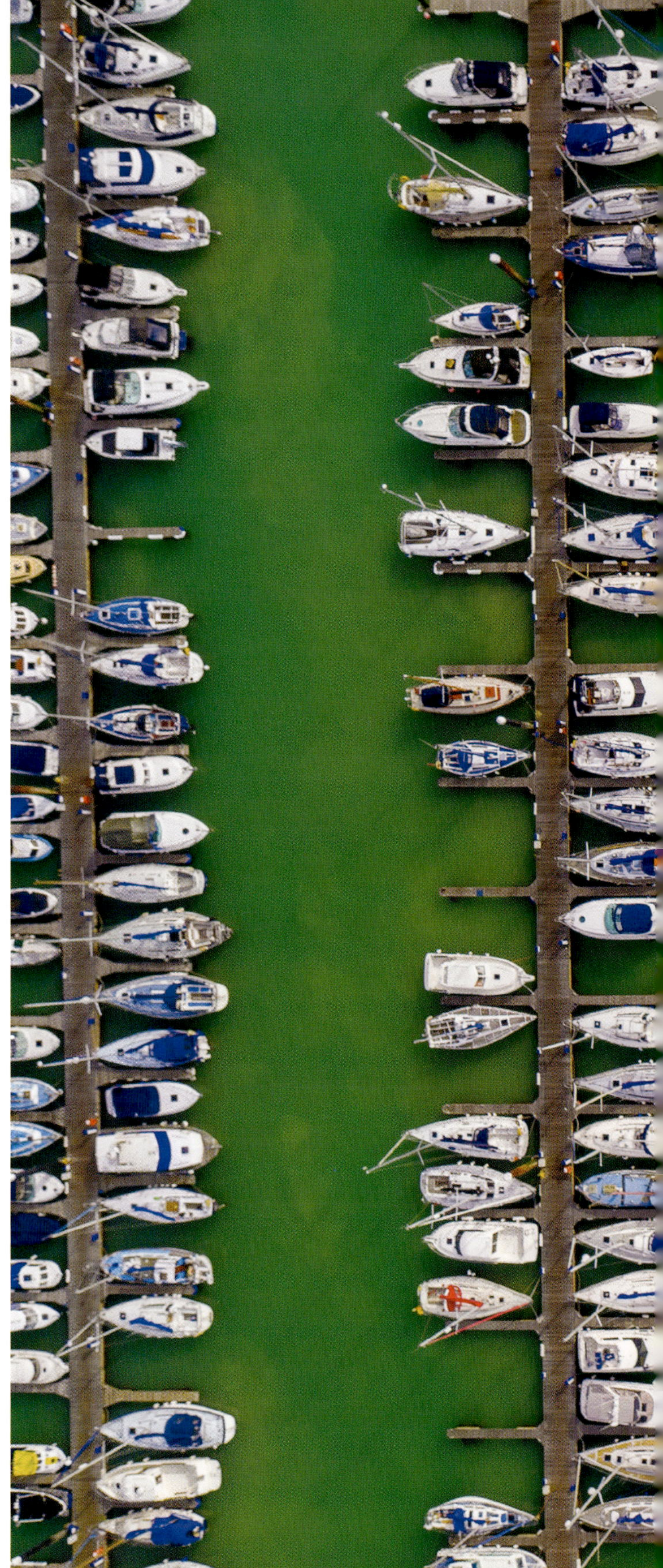

▶ The general public is rightly concerned about privacy issues involving drones. In most countries it's against the law to photograph people when they have a reasonable expectation of privacy. However, most drones are equipped with wide-angle lenses, so from 400ft (120m) as in this shot, people are not going to be recognizable.

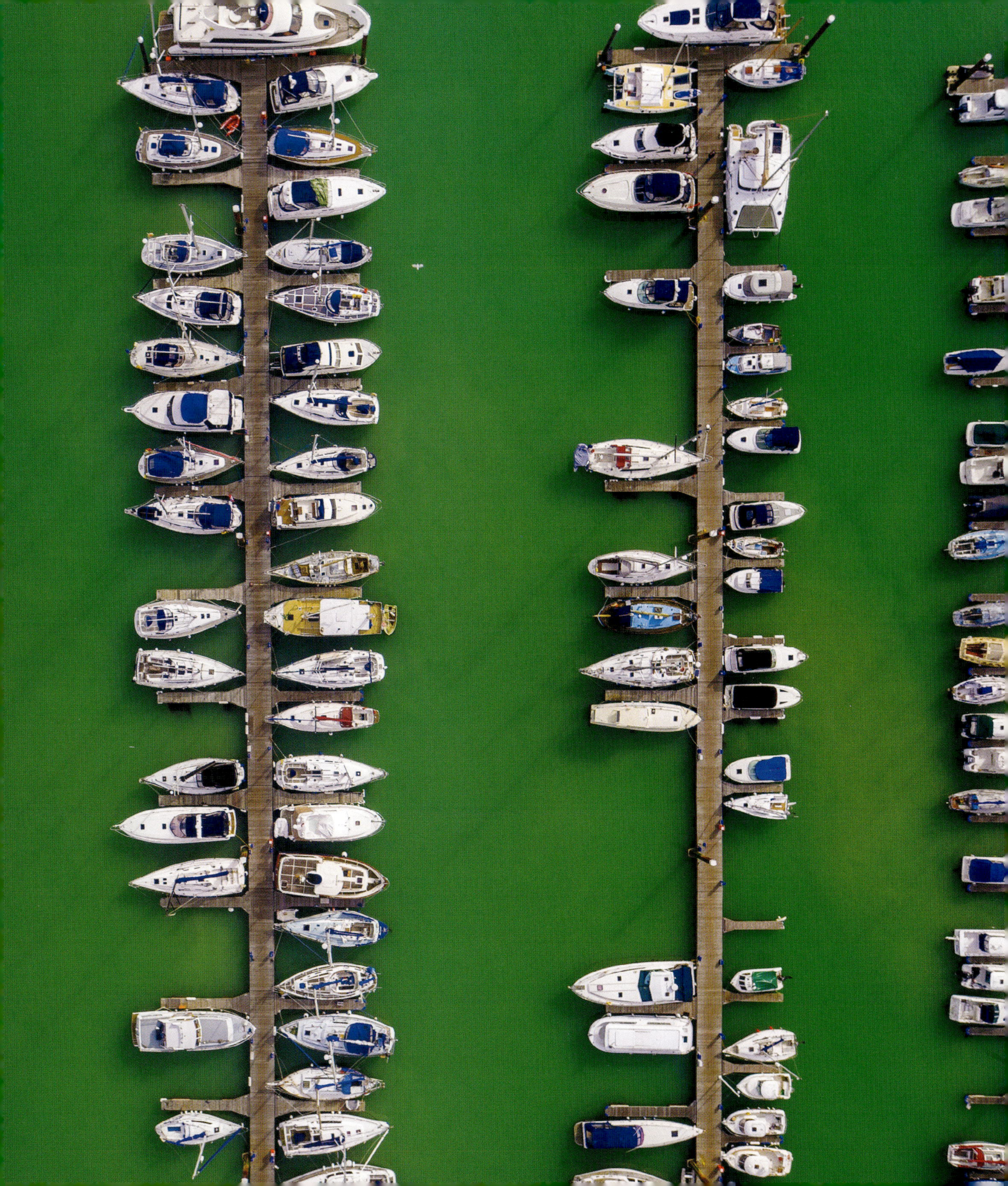

WEATHER

All drone pilots need to have a keen awareness of both the current weather situation and the weather forecast. People who habitually spend a lot of time involved in outdoor activities may already have an instinctive feel for the meteorological conditions, but it is worth appreciating how they affect drones.

WIND

One of the most obvious weather-related issues that will affect your drone-flying activities is the wind. Your drone is likely to have a maximum speed indicated in its list of specifications. It's important to realize that this is an air speed, in other words it's the maximum speed it can reach relative to the air around it. So, if your drone's maximum speed is 25mph (40kmph) and you are in a steady 25mph (40kmph) wind, your maximum ground speed—heading into the wind—will be 0mph/kmph. Of course, if you headed downwind, you might hit 50mph (80kmph) over the ground, but you're unlikely to get your drone back!

The simple rule of thumb is that you shouldn't fly your drone in a wind that is anywhere near its maximum speed. For most GPS-equipped outdoor drones, this might mean only flying in winds below around 18mph (30kmph). However, remember that when you're standing at ground level you may well be sheltered from most of the wind by buildings or trees. Wind speed tends to increase as you gain height, so look at the branches of the highest trees around to see if they are moving significantly—if they are, proceed with caution when gaining height.

If you get into wind that is significantly above the top speed of the drone, the drone's GPS hold will not be able to maintain the drone's position and it will start to drift away with the wind. Should this ever happen to you, the best thing you can do is to get the drone down as quickly and safely as you can. Once you are closer to the ground you may find the reduction in wind speed allows you to fly back at a lower altitude.

It is not just wind speed that will affect your drone, though, as turbulence is also a big issue. Wind flow will be disrupted close to obstacles: upwind of a large obstacle the wind will have to rise to get over it, but on the downwind side there may be severe turbulence—confused air moving in a variety of different directions, often with strong downdrafts. This may start becoming noticeable in winds above 10mph (16kmph) and it could be really problematic in winds of 20mph (32kmph) and above. Consequently, you should be very wary of the downwind side of large obstacles in stronger winds. This problem can also be compounded by the reduced GPS signal close to large obstacles, which makes the drone even less controllable.

TURBULENCE

I have had first-hand experience of strong turbulence on a number of occasions. Perhaps the most striking was an early morning shoot on some coastal cliff tops. I had checked the forecast and there was a manageable 12mph (20kmph) wind speed forecast for that time. However, when we got to the location, it was closer to 20–25mph (around 30–40kmph), due to our cliff-top position. As my drone was capable of 30mph (48kmph), these were flyable conditions for that particular drone—it was marginal, but I decided to go ahead.

The take-off and landing spot was about 12ft (4m) downwind of the edge of a vertical cliff and the wind was blowing directly onshore. What I found was that the first 12–15ft (4–5m) above ground level were incredibly turbulent. The drone would be hit by unpredictable downdrafts, and suddenly lose up to a yard of altitude. This effect is called "rotor" and well known by paraglider pilots who take the risks very seriously. Once clear of the turbulent zone, the drone flew steadily with no problems: I just had to make sure that I spent as little time as possible close to the ground and got clear and upwind of the cliff edge as quickly as possible.

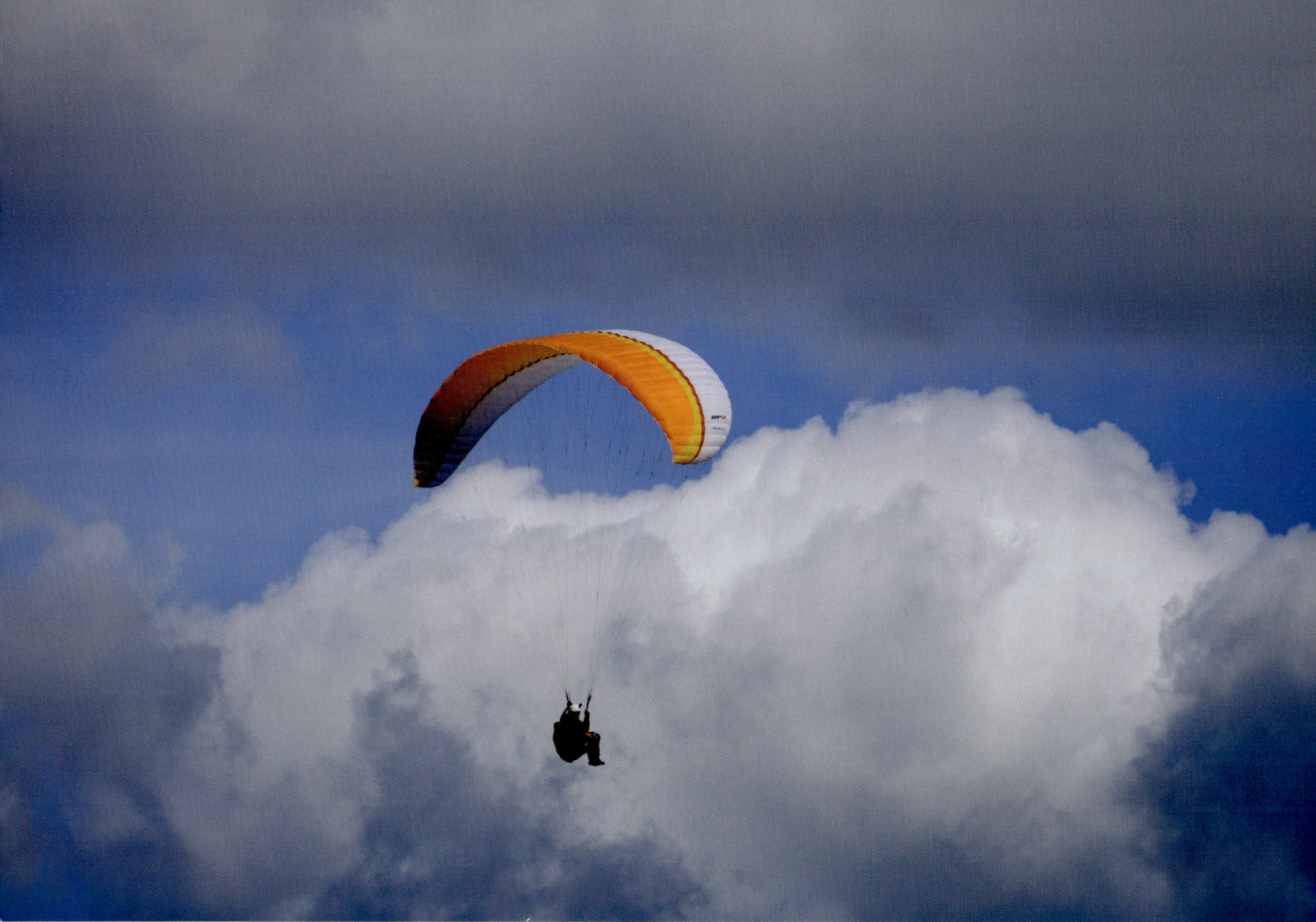

PRECIPITATION

Water and electrical components do not play nicely together, so it is a good idea to avoid flying in any sort of rain or snow. However, don't panic if it starts raining while you are flying. Simply bring the drone in as quickly as you safely can and get it under cover. It's happened to me several times, with no lasting damage.

There are now a few drone models on the market that claim to be weatherproof/waterproof, so these may be worth considering if you really need this feature. However, rainy conditions often don't provide you with the best photographic opportunities, so even if you prevent water getting onto your electrical components, you will likely get water drops on the camera lens that ruin your shots.

FOG

Flying in fog is not recommended, as you should always have a clear line of sight to your drone: not only do you risk losing your drone, but the risk of collision with obstacles or other aircraft is increased. As with precipitation, dense fog will also condense on your drone and camera lens, and may interfere with its function.

If you want to get great shots above the clouds or fog, try to find a location where you can take off above it, maintaining a clear line of sight to the drone.

▲ Steer clear of fog and clouds. Not only are there visibility implications, but larger clouds, such as developing cumulus clouds, often bring significant turbulence with them.

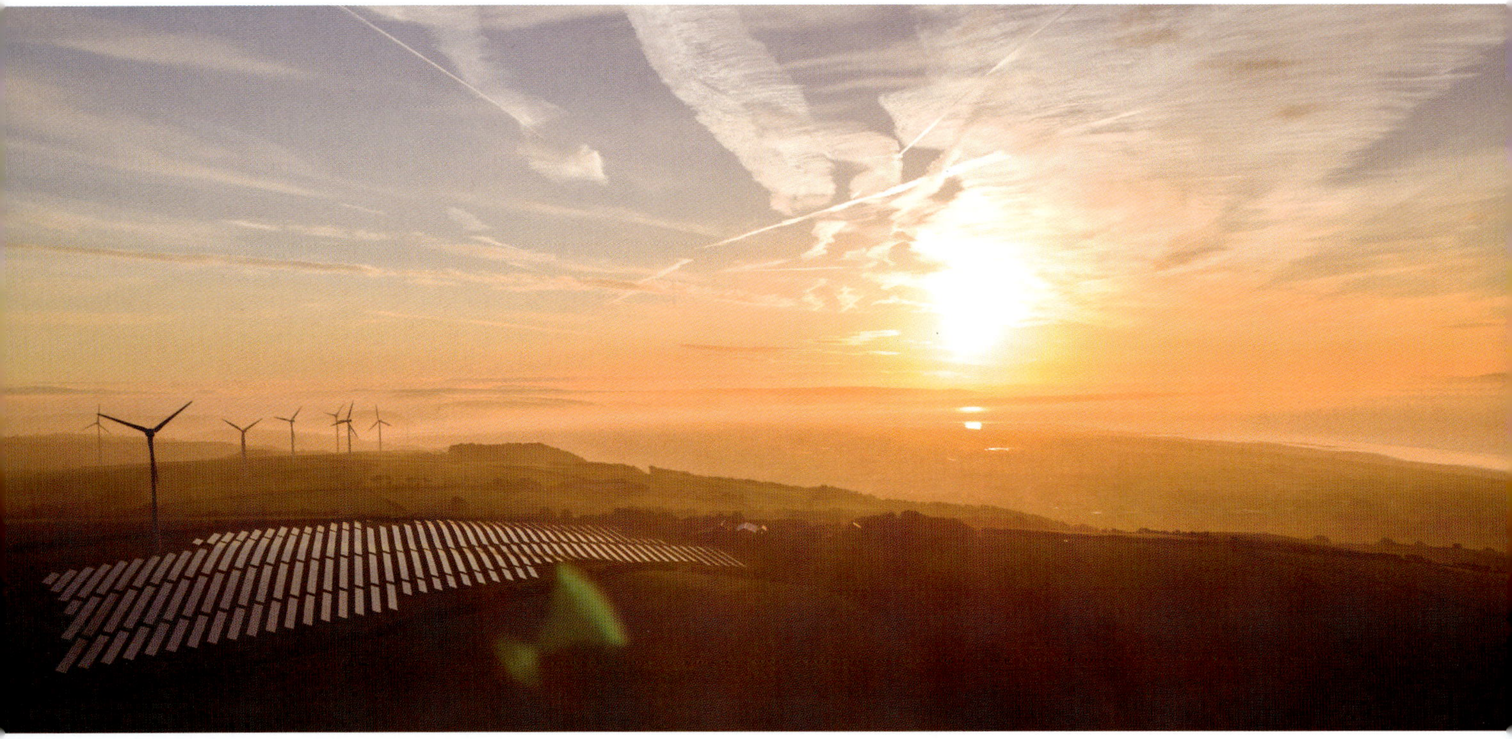

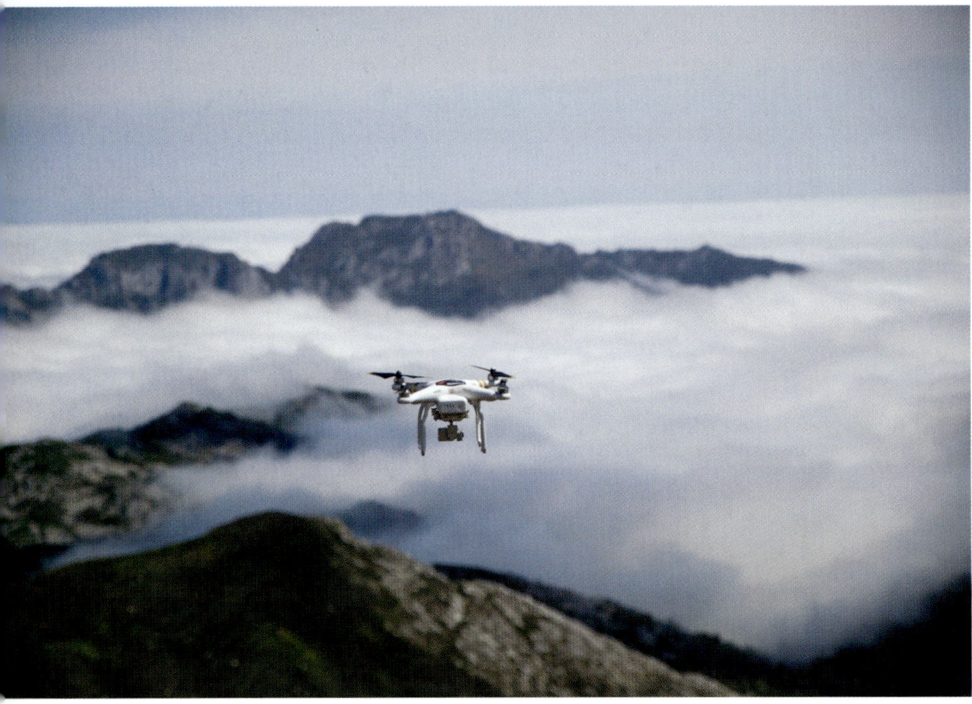

▲ Sunrise sometimes means light winds and an attractive layer of thin mist, so it's worth getting out of bed early!

◀ Flying in foggy conditions can make for some epic shots, but always ensure you can see the drone, preferably by choosing a take-off spot above the fog layer.

TEMPERATURE

You can usually find the recommended operating temperature range for your drone in its specification sheet—this is generally in the region of 32–104°F (0–40°C). However, this is for the electronics. Lithium polymer batteries can be more sensitive to colder temperatures and do not operate optimally below temperatures around 60°F (15°C). This is not as much of a problem as it may first appear, though, as once the drone is flying the batteries warm up significantly.

What it does mean is that if it's cold, you should keep the batteries warm right up until take off, either in a heated car, in an inside pocket under a coat, or in a dedicated battery heater. Some models of drone will give you a warning in the app and not allow you to fly if the battery is too cold. They may also be able to do a preflight warm up, depending on the model.

AIRBORNE DUST & SAND

In arid environments—particularly if it's windy—be wary of airborne dust and sand, which can be highly destructive to moving parts such as bearings and motors (pilots who habitually fly drones in such environments will change electric motors and bearings frequently). It may also reduce the range of your control frequency and video link.

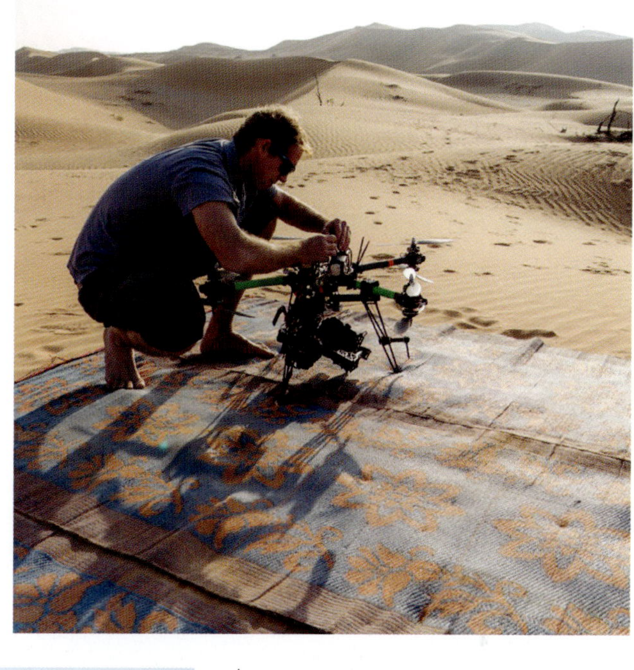

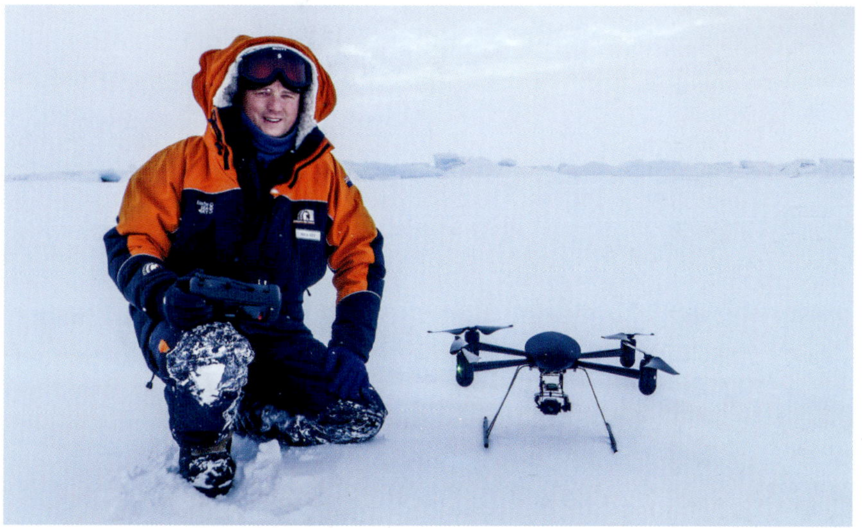

▲ Drone pilot Miguel Willis prepares his aircraft in the sands of Arabia's Empty Quarter (or "Rub Al Khali"). In conditions such as this, special care must be taken to minimize the ingress of sand and dust into motors and other moving parts. You can see more of Miguel's work on pages 70–71.

◀ When operating a drone in very cold temperatures, care must be taken to keep the battery warm right up until take off. A temperature of around 68°F (20°C) is ideal.

PLANNING YOUR MISSION

There are many things that can conceivably go wrong on a drone outing, but the planning you do before you even get to the location can be the difference between a success and a catastrophe. Here we'll look at how you can apply the safety guidelines to a drone shoot to minimize any risk.

To start with, it is a good idea to have a checklist for the location you are operating in, which might look like this:

- Do you need permission from the landowner?
- Check the weather forecast for wind speed and direction, the possibility of rain, and the temperature.
- Airspace: is there sufficient clearance from nearby airports?
- Is there the required clearance from buildings, roads, and the public (note that this distance may increase if using a larger drone and/or flying near large gatherings of people)?
- Check access and be aware of public rights of way in the vicinity of the flight.
- Is there sufficient distance from obstacles at the take-off/ landing spot (typically 5–10 yards/meters)?
- Are there any other obstacles you need to be aware of, such as trees, powerlines, buildings, and so on.
- Are there any potential sources of interference, such as radio transmitter masts or strong Wi-Fi activity?
- What is your alternative landing spot?

LANDOWNER PERMISSION

Even if you are flying from your own land, if there are other buildings close by (usually within 50 yards/meters) you should inform the owners and ask for their permission before flying. If the land you are flying from is privately owned you should also ask for the landowner's permission.

On public land, where there are rights of way, you might well be OK if you are flying a small drone recreationally and maintain the recommended clearance from people and buildings. However, if it is commercial work you will need permission.

This may sound like a lot of work, but often it is relatively painless. More often than not, landowners and farmers will be happy to let you fly, just as long as you're not disturbing anyone—you can even offer them a free aerial photo of their property as a "thank you."

◀ Particular care should be taken when planning flights in or close to residential areas. You need to have permission from the owners of any buildings within a 165ft (50m) radius of your flight and make sure you have control over anyone who might move within that radius. If the area is congested you'll need to increase the "safe" distance according to the local regulations.

GOOGLE EARTH

You can scout potential locations on Google Earth, which will help you determine how feasible it is to fly there. To start with, you can download Google Earth format 3D airspace maps or look at a separate map, which will tell you if you are operating in restricted airspace or there are any other aviation hazards.

For current announcements on aviation-related matters, you can also plan a flight with *www.skydemonlight.com*, which will give you a list of current NOTAMs (NOtice To AirMen). This should inform you of any unusual airspace activities on the day of your flight, such as airshows or scheduled military activity. Some websites and apps (such as NATS Drone Assist) not only detail restricted airspace and current NOTAMs, but also allow you to announce your flight so other air users are aware of your activity.

Once you have checked the airspace situation, use Google Earth or Google Maps to take a good look at the surroundings. Remember you should select a take-off spot with sufficient clearance in all directions from people, buildings, and roads: once you are airborne you also need to maintain a safe distance from congested areas, so check that too.

OBSTACLES

It may seem obvious, but having a good mental map of the trees, telegraph poles, and power lines can really help if you need to deal with an unexpected scenario. In conjunction with a detailed weather forecast you will also be able to determine if there are any areas you might want to avoid due to possible wind turbulence. Bear in mind any seasonal hazards as well—the low winter sun can make it hard to see your drone in a particular direction, for example.

RISK ASSESSMENT

As a recreational pilot there is no obligation for you to carry out a risk assessment prior to each flight. Nonetheless it is worth at the very least considering what the risks are and how you might minimize them.

If you are flying commercially you will often be required to complete a detailed risk assessment prior to each job. This my be required for your client's insurance to be valid. The grid below details a selection of common risks.

RISK	MITIGATION
Collision with obstacles	• Ensure sufficient pilot training and experience • Use a drone with built-in obstacle avoidance • Have an observer on watch
Collision with public	• Establish a 100ft (30m) exclusion zone around take-off and landing point • Observe minimum safe distance during flight • Have marshals in place to temporarily limit public access to operations area
Collision with other aircraft	• Stay out of restricted airspace • Fly below 400ft (120m) • Have an observer on watch in addition to the pilot • Be aware of possible incursions of low-flying aircraft and be prepared to land
Lithium polymer battery fires	• Good battery care • Charge in fireproof containers • Check batteries for damage
Loss of control frequency	• Set automatic return to home (RTH) and land

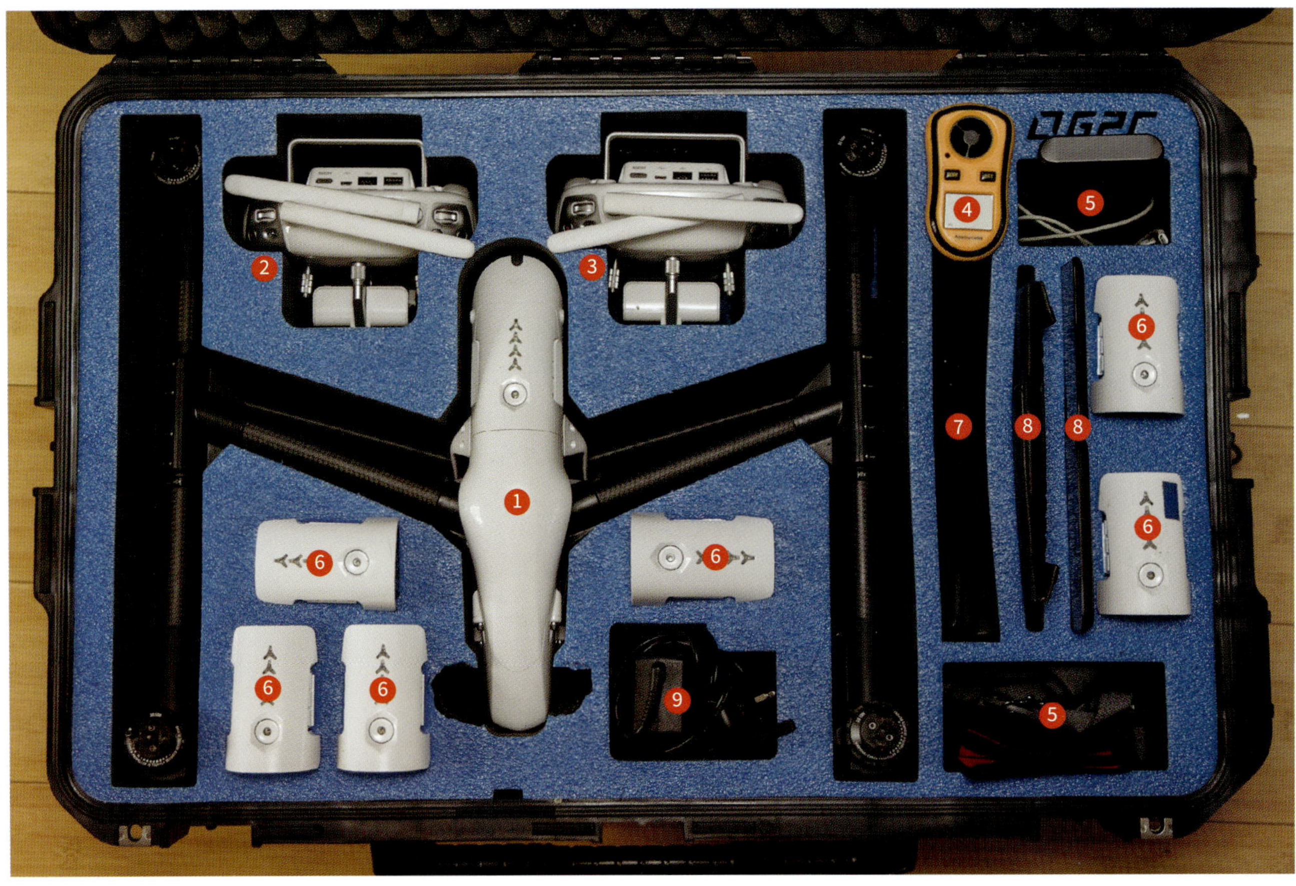

▲ A well laid out case can go a long way to helping you make sure you don't forget anything vital. It's worth investing in a good quality, durable case.

PACKING CHECKLIST

There are few things more annoying than turning up to a new shoot location with perfect weather, getting out your drone, and finding you've forgotten to pack a tiny, yet vital piece of equipment. To avoid this, it's worth having a checklist of what you need to pack and ticking it all off the day before your flight. Obviously, what you need will vary enormously according to your setup, but things that can get forgotten regularly include the list at the right:

1. Drone
2. Aircraft remote control unit
3. Gimbal remote control unit
4. Anemometer
5. Cables & straps
6. Battery
7. Propellers
8. Tablets (iPads)
9. Charger

▶ A last minute preflight check for any obvious mechanical or electrical anomalies can help prevent mishaps.

PREFLIGHT CHECKLIST

It may seem like a waste of time (particularly once you become an experienced pilot), but a few minutes spent running through a checklist before a flight can catch issues before they develop into something more serious. Even seasoned drone pilots can overlook crucial details in the rush to get a shot.

- **Remote control**: Assemble the RC unit, with tablet and cable.
- **Remote control battery**: Should be at least 75%.
- **Aircraft**: Check aircraft and fit gimbal; remove gimbal lock.
- **Antenna**: Check antenna are fitted/in correct positions.
- **Propellers**: Check they are fitted correctly and firmly locked.
- **RC switches**: Correct flight mode and landing gear down.
- **Tablet**: Check the app is working correctly and you have sufficient battery charge.
- **Aircraft**: Check obstacle clearance and orientation.
- **Power on**: Check flight battery is at (or near) 100%.
- **Compass**: Check compass is in the range 1400–1599.
- **IMU**: Check IMU is within ±0.1 variation.
- **Failsafe**: Set an appropriate failsafe height.
- **GPS**: Ensure you have a good satellite lock (ideally seven or more satellites).
- **Flight mode**: Check correct mode selected.
- **Start motors**: Using CSC command.
- **Home point**: Check the home point location is correct.
- **Motor startup**: Ensure motors sound correct before take off.
- **Ascend**: Bring the drone to hover approximately 2 yards/meters above the ground.
- **Control tests**: Test all controls.
- **GPS & ATTI test**: Test both modes are functioning correctly.
- **Landing gear**: Raise landing gear (if applicable).

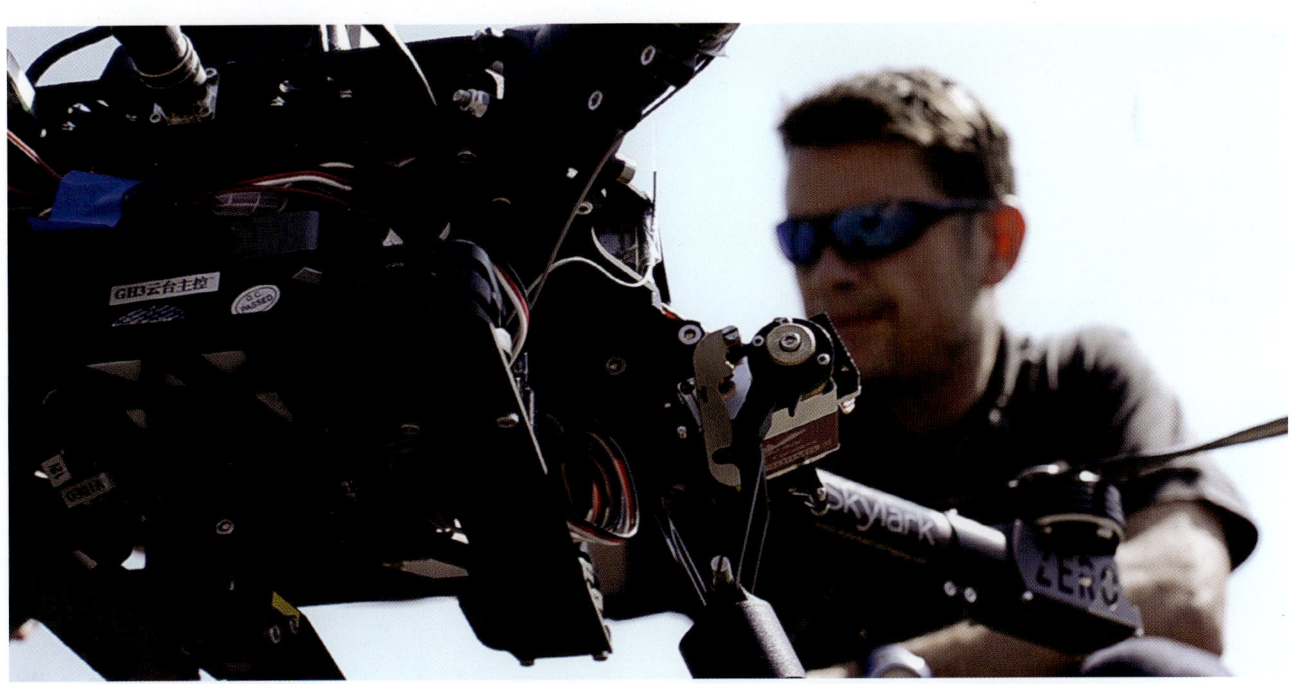

PROFILE: MIGUEL WILLIS

Miguel Willis is an Australian freelance cameraman and drone operator who specializes in action sports, adventure, and wildlife photography. He has shot in a wide variety of locations around the world, with these images taken during an expedition across Arabia's notorious Empty Quarter.

The expedition had plenty of challenges, as Miguel explains: "In the Empty Quarter the fine sand gets absolutely everywhere, so you have to be really careful to protect the drone's delicate mechanics and electronics. Yet despite the logistical and practical difficulties, drones proved a fantastic platform from which to showcase the vastness and isolation of one of the world's most spectacular wilderness areas."

▶ A simple top-down composition allows a camel-train to cast pleasing shadows as the sun gets lower in the sky.

▼ Dwarfed by the immensity of Arabia's Empty Quarter, a small group of travelers pick their way through the dunes. As with the image opposite, this is a frame grab from a 4K video, shot using a custom-built X8 drone fitted with a Panasonic GH4 Micro Four Thirds camera and Olympus 12mm f/2 lens.

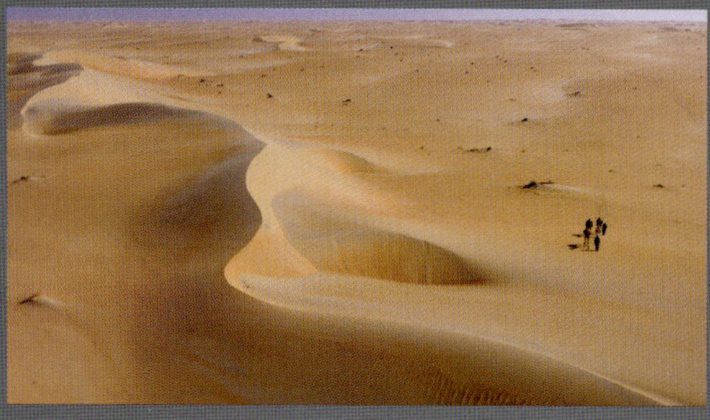

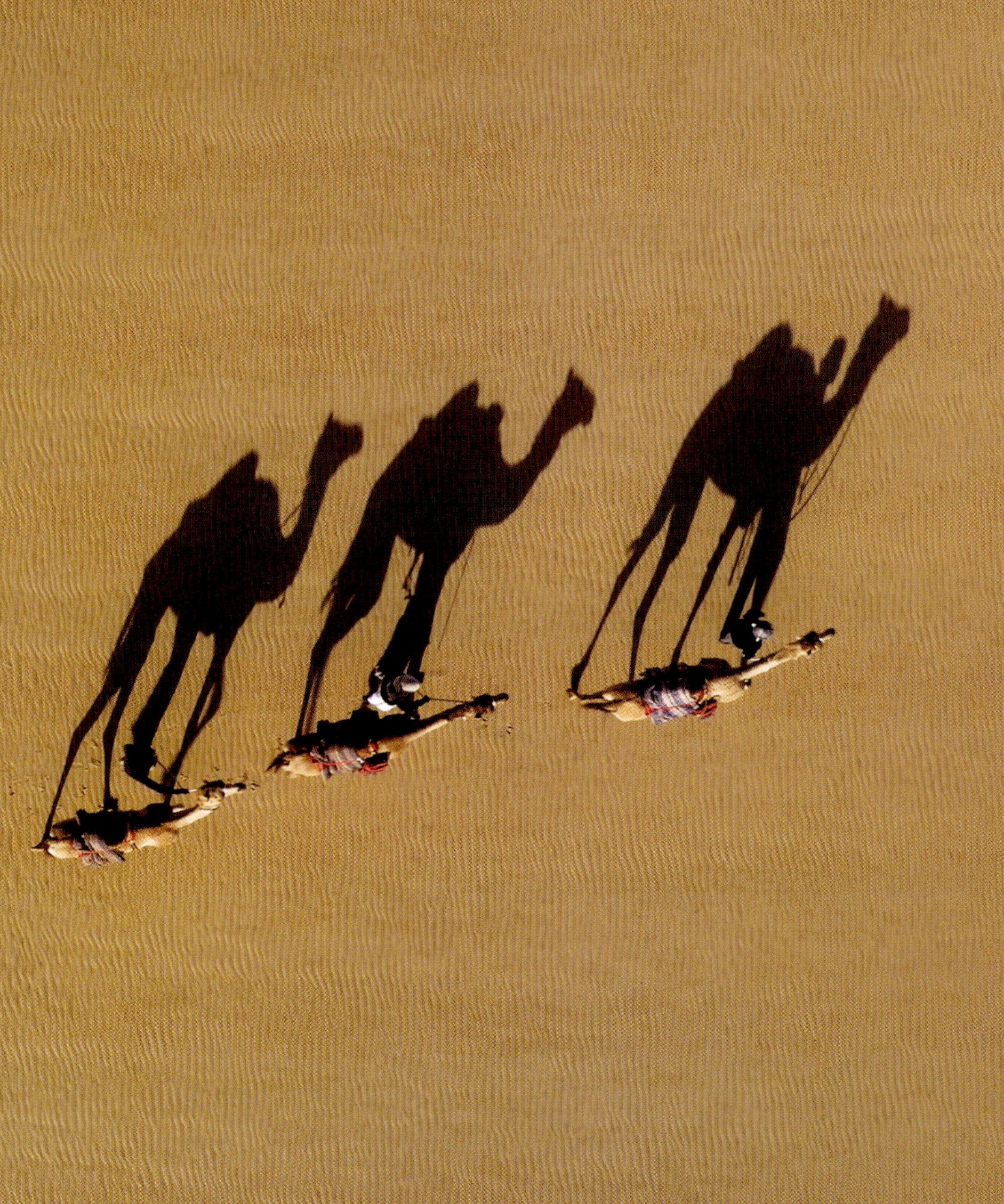

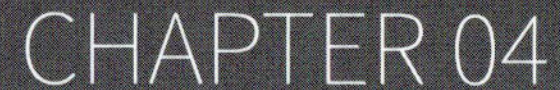

CHAPTER 04

DRONE SETUP & FLYING SKILLS

As a photographer you might be very familiar with the way that cameras work, but setting up a drone correctly and flying it safely and precisely requires thought and a good amount of practice. Provided you follow a few simple guidelines, don't try to rush it, and always err on the side of caution, getting to grips with this new technology should be both enjoyable and rewarding.

This chapter outlines how you can familiarize yourself with the operation of your drone prior to your first flight, as well as exploring the various flight modes you might be able to use, and some simple exercises that will help you hone your flying skills.

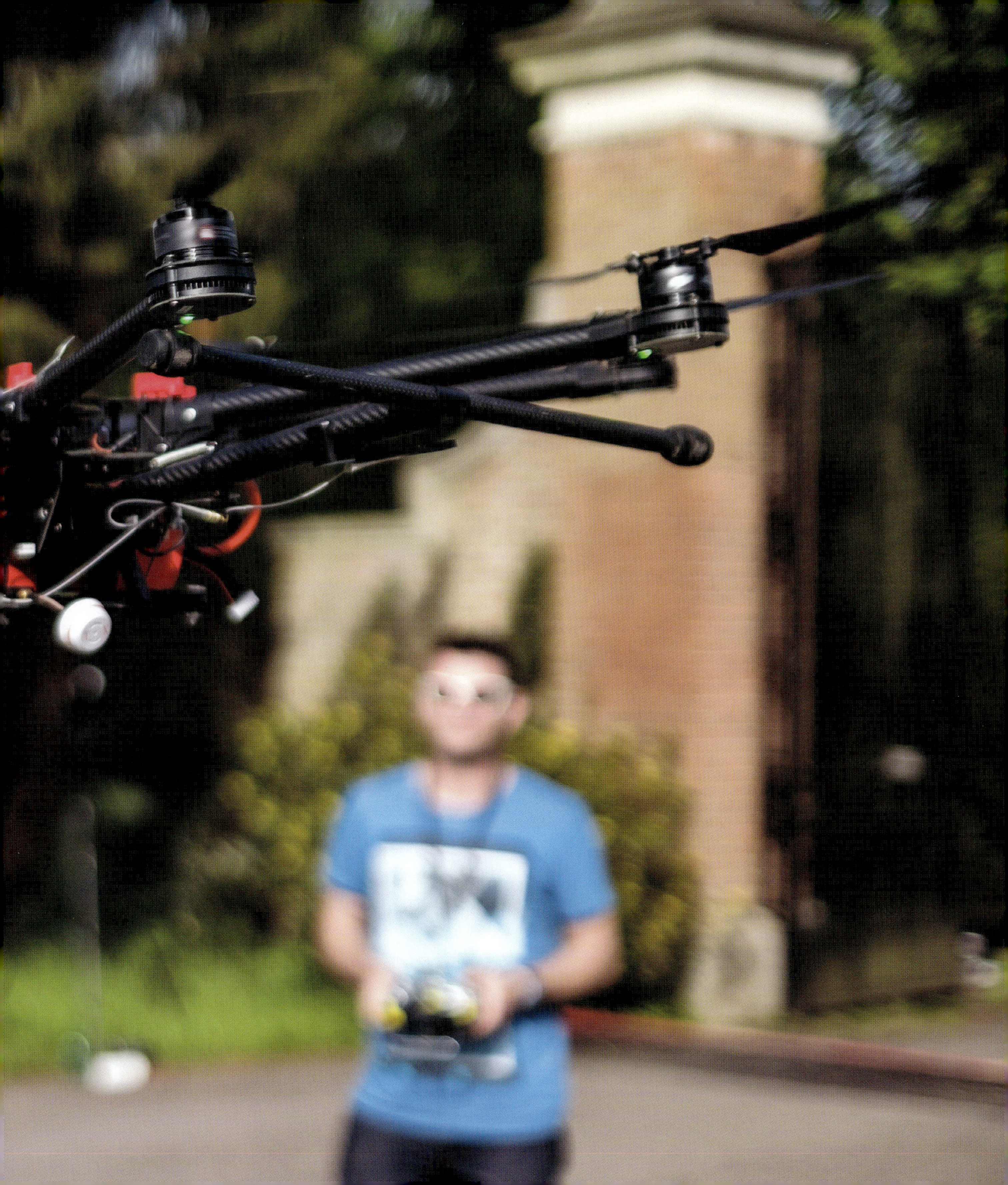

GET TO KNOW THE APP

Even if you have a relatively simple drone there is a number of basic steps that could save you from a costly or painful accident. These are not all things you necessarily need to do on every flight, but you should definitely bear them in mind if you want to avoid any in-flight surprises. A little time spent familiarizing yourself with both the physical workings of the drone and its app or software can only help you to avoid mishaps and deal with unexpected events.

Most drones in the mid-sized camera drone category are either designed to be used in conjunction with a smartphone app or have their own built-in display running an app. Before you take off you should familiarize yourself with how to use the app, which is something you can do indoors—you can safely switch on the transmitter and the drone, but leave the propellers off as a safety precaution.

Always follow the user manual carefully, as it's often important to do things in a certain order when setting up. To start with, your phone or tablet will possibly need to be connected to the remote control transmitter via a USB cable. Then the usual order is to switch on the transmitter and then power on the drone. However, you should make sure you have removed any gimbal lock before powering on, or you risk damaging the gimbal.

Once the drone and remote control are switched on, you may find that the app launches automatically (if it doesn't, you will have to launch it yourself). Either way, the app should detect the transmitter, confirm the drone is connected, and provide you with an on-screen view through the drone's camera. If you don't get a connection first time, unplugging and reconnecting the USB cable can often resolve the problem.

▼ The "Compass Calibration Dance" usually involves turning through 360 degrees while holding the drone horizontally...

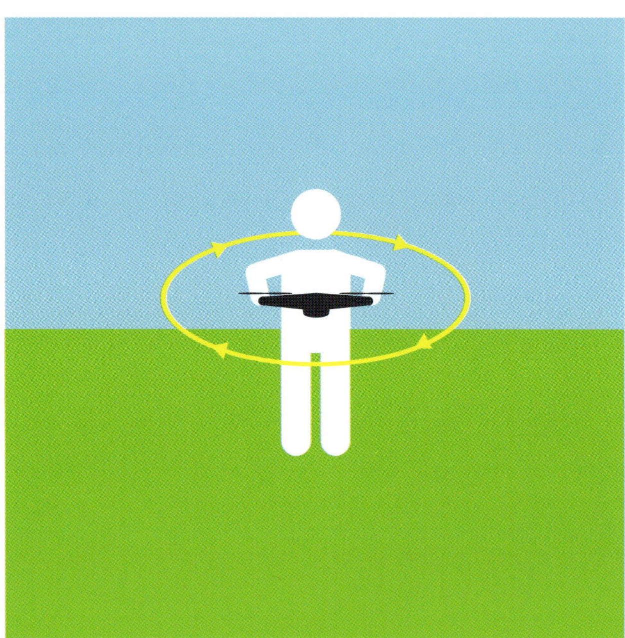

▼ ...and then turning through 360 degrees with the drone held vertically.

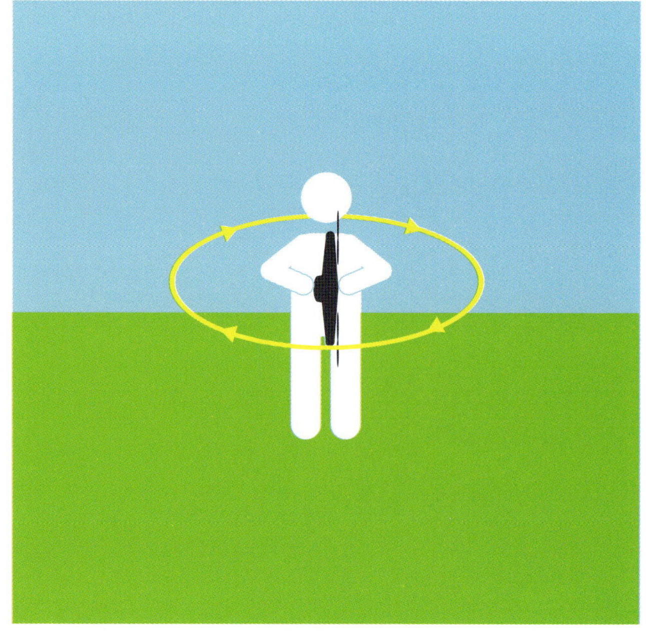

Depending on the model of drone you own, you will be able to read off some or all of the following telemetry on your display (telemetry is the transmission of data from your drone):

- Flight time remaining
- Battery voltage/percentage
- Number of GPS satellites
- Height above take off
- Horizontal distance from take off
- Flight mode
- Camera settings
- Aircraft heading
- Horizontal speed
- Vertical speed
- Control signal strength
- Video signal strength

In addition, you may get various warning alerts, such as low battery, cold temperature, propulsion limitations, and GPS signal lost, amongst others.

The app will also allow you to change a wide array of settings, both drone related and camera related. It's important that you familiarize yourself with all of this, because if you have to change anything while the drone is flying, you don't want to spend too long searching for the right icon or scrolling through menus.

Once you've digested the information in the app, and determined how you can change important settings, there are several procedures you should become familiar with in order to minimize the chances of a mishap:

Compass calibration: Your drone will probably have an electronic compass built in. In general, it's not necessary to calibrate the compass before each flight, unless you are in a markedly different location from the previous flight (by which I mean hundreds of miles from your usual flying area). You should be careful to perform the calibration in an open area, well away from large metal objects (including underground ones, such as pipes and cables), or anything magnetic that might be close.

The compass calibration procedure will vary slightly according to your drone, but it usually involves activating a calibration mode—either on an app or with some switch toggling on the remote control—and performing a 360-degree turn whilst holding the drone firstly in horizontal orientation and secondly in vertical orientation (without propellers attached!). You should get a confirmation of success involving changed LED status on the drone and a confirmation in the app. If calibration fails you should repeat the steps until you get a successful calibration.

IMU calibration: The IMU (Inertial Measurement Unit) uses a combination of accelerometer, gyroscope, thermometer, and barometer to measure and report the specific force, angular acceleration, and attitude of the drone. Most of the time it can be left alone to do its job, but there are times when it might be helpful to recalibrate it. It's a good idea to do an IMU calibration after a firmware update, for example, or after the drone has been repaired following a crash or exposure to other strong forces. You may also be prompted to calibrate the IMU by the app on your phone or tablet if any abnormalities are detected.

IMU calibration should be carried out in a cool location and when the drone is cool, so it shouldn't have been flown for the previous 30 minutes or so. You should ensure the drone is level and resting on a stable surface, such as a sturdy table. Use a spirit level to get the level exact, placing sheets of paper under each foot until the drone is perfectly level. Calibration should then be initiated using the app or software and the drone should not be touched or moved until it is complete.

▼ **You can explore the flight modes and camera functions of your drone's app in the comfort of your own home.**

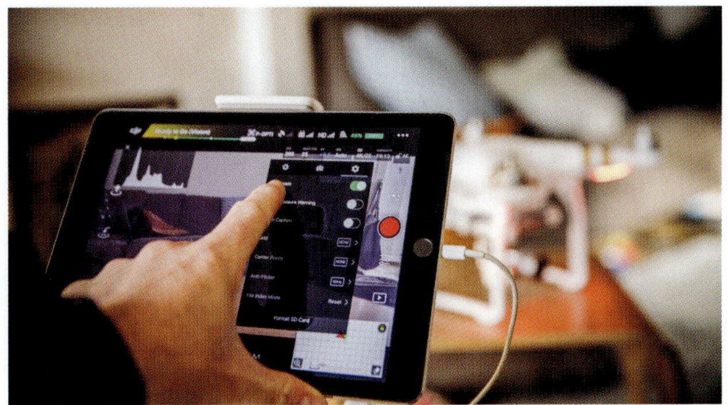

Check sensor readings: Whether or not you have a calibrated compass or IMU, you should check your sensor readings on every flight if your flight app allows it. On some of the newer DJI drones this is found under *Main Controller Settings > Advanced Settings > Sensors*.

Here you can check the Mod values of the Gyroscope, Acceleration, and Compass: the value should be very close to 0.00 for Gyroscope and close to 1.00 for Acceleration when the drone is stationary on the ground. Any significant deviation from these values means you need to calibrate the IMU.

For Compass, the value will be shifting, even at rest, but should be somewhere between 1400 and 1599. If not, you need to calibrate the compass.

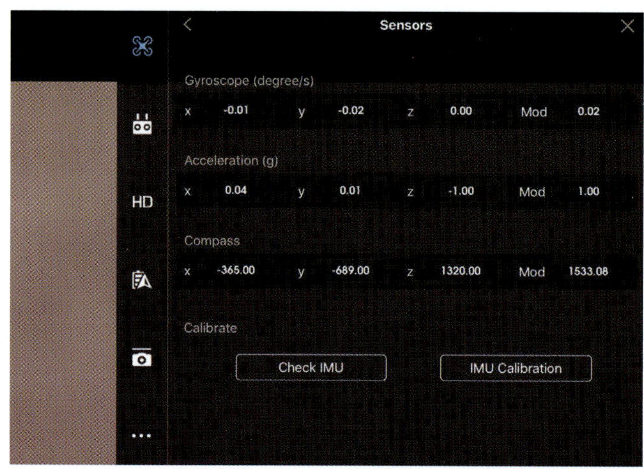

Battery: Ensure your battery is fully charged at the start of each flight. Starting a flight with a partially charged battery may prompt misleading battery status data in flight. You should also plan to land your drone at 30% battery capacity, as discharging a lithium polymer battery fully will damage it, often permanently. Having some reserve battery power at the end also gives you the flexibility to make last minute changes to your landing plan (if, for example, your planned landing spot is occupied).

DJI drones have a "Smart Go Home" feature, where the drone will automatically fly to the home point and land when the battery level hits 10%. If this feature is disabled, it will simply land at its current position. There is no correct setting for this; the most suitable option depends on your flying environment.

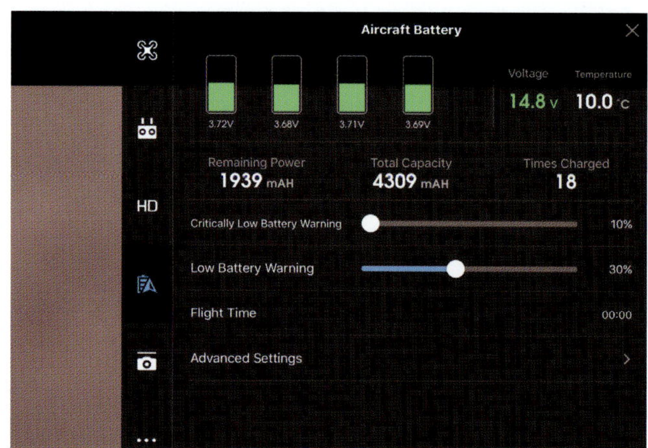

Check your failsafe settings: Almost all modern drones have the option to return to their take-off spot if communication with the remote controller is lost. Usually, the default setting is that once communication is lost, the drone will ascend or descend to an altitude of 65ft (20m), fly to the point directly above its take-off spot, and come in for a vertical landing.

You can adjust these parameters in the app, so think about the height and position of any obstacles in the area you are flying. Also bear in mind that the GPS positioning may only be accurate to around 5–10 yards/meters, so you need to make sure there is a clear radius around your take-off spot in case of an unplanned return to home.

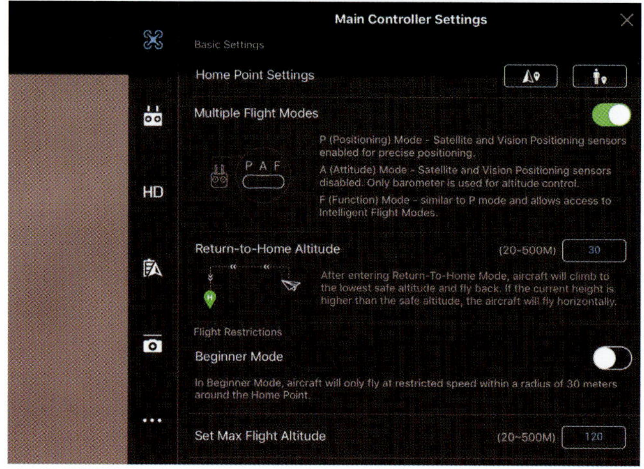

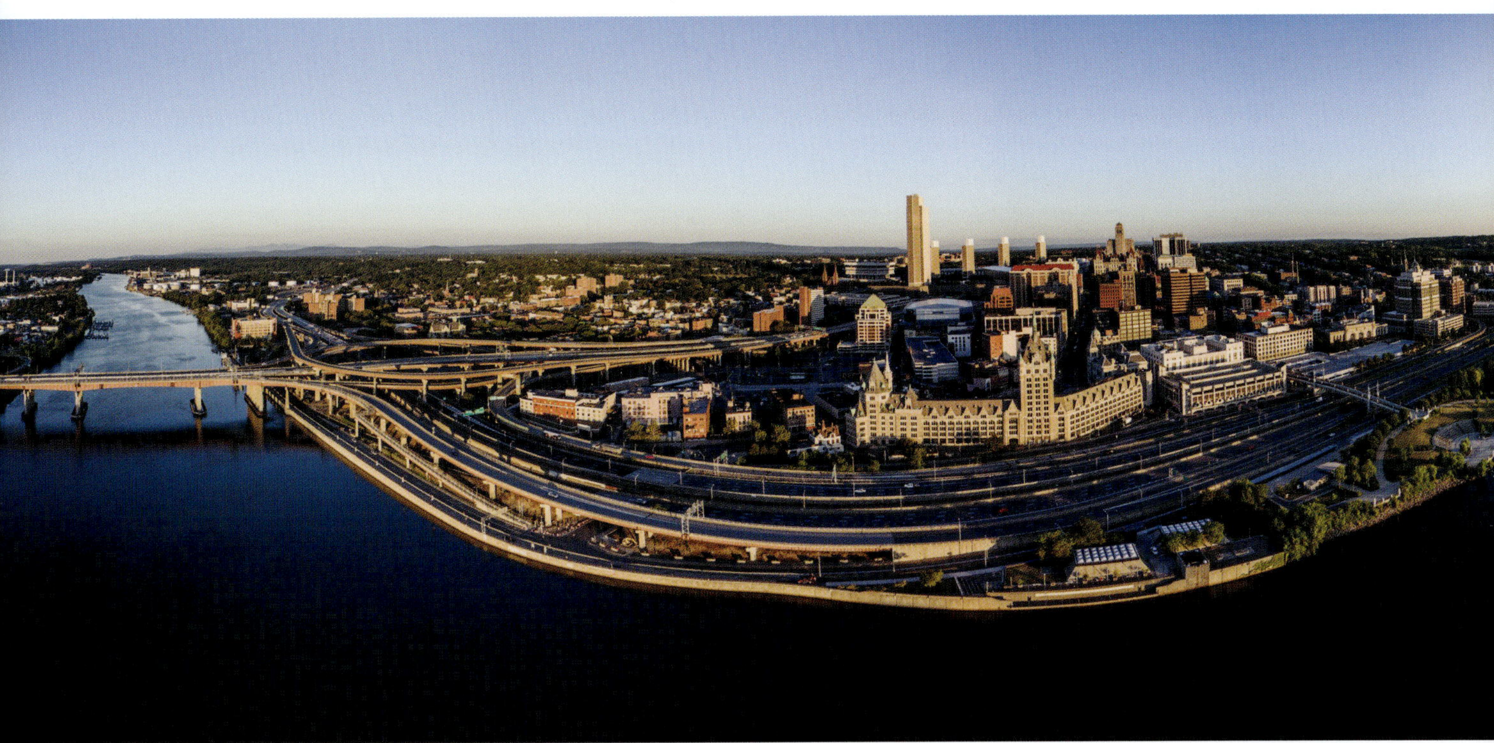

Finally, just prior to take-off, the drone will report that the home point has been updated and will suggest you check it on the map in your app. It's easy to skip this, but it is worth doing, just to ensure that the home point is correct.

Phone & tablet to airplane mode: Many drones are controlled using the 2.4 GHz frequency, which is the same frequency as Wi-Fi. Although phones and tablets should be able to avoid interference, to be on the safe side—and maximize range—it's wise to switch your phones and tablets to airplane mode.

Be prepared to fly in ATTI mode: Your drone's GPS abilities are great at helping to keep it stable in the wind and enabling various "smart" flight modes, but it's not infallible. Any sort of overhead obstruction, such as dense tree cover, foliage, or tall buildings will compromise the GPS signal. Particularly strong solar storms could also seriously degrade the GPS satellite accuracy.

For these reasons you should be prepared to fly the drone without the help of GPS mode. Firstly you may need to enable non-GPS mode. Then you can use the mode switch on your remote to select A (ATTI or Attitude Mode). Give yourself plenty of time and space to practice flying in this mode, as the drone may respond faster and move more quickly than usual. It will also drift with the wind when there is no control input.

When you push the drone in a direction, it will keep going in that direction until either you or the wind stops or deviates it, so you should always be aware of the wind direction and strength. Imagine and mentally plan how you would get the drone back in the event of an unexpected failure in GPS lock. In the event of sudden and unexplained movement of the drone in flight your first course of action should be switching to ATTI mode to regain control.

▲ You should practice using ATTI mode so you are prepared for any sudden GPS signal loss or compass errors. It is best to practice on a day with light winds, staying well away from buildings and other obstacles.

Proximity sensors & obstacle avoidance: Every year drones are becoming "smarter," with an ever-increasing array of technologies that makes flying them easier and safer. One such technology involves downward-facing cameras that detect patterns on the ground and use these to stabilize the drone in the absence of a strong GPS signal—when flying indoors, for example. However, this typically only works at very low altitudes (below 10 yards/meters) and can be fooled in low light, with repeating patterns or very homogenous surfaces, and over water. In such circumstances it's often safer to disable this feature.

Similarly, some drones have built-in obstacle avoidance that uses one or more pairs of small stereoscopic cameras to detect possible obstacles and either stop the drone or navigate it around the blockage. This is a great feature to have, but you should know its limitations and avoid relying on it.

One common error is to assume that your obstacle-avoiding drone can see in every direction. Several models only have forward-facing obstacle avoidance, which will not help you if your drone is flying sideways or backward. Furthermore, the obstacle avoidance may not detect very fine obstacles such as overhead cables or very thin tree branches. It is also automatically disabled in some flying modes (such as Sport Mode).

When flying very close to the ground (under 10 yards/meters), your drone may have an acoustic altimeter that enables it to follow the terrain, gently rising or descending to maintain a steady altitude above the ground. This is usually a great feature to have, but there are times when it can be problematic. For example, I was once flying over some outdoor steps that had a low overhanging tree above them. As the drone passed over the steps it automatically gained altitude and it was only some quick reactions on the stick that stopped it from hitting the branches overhead. Again, when circumstances demand, be prepared to disable the sensors and take full control of your flying.

▼ **Many newer drones have an array of sensors to detect the proximity of nearby obstacles or the ground. These should be treated as a safeguard, but you shouldn't rely on them to prevent all collisions.**

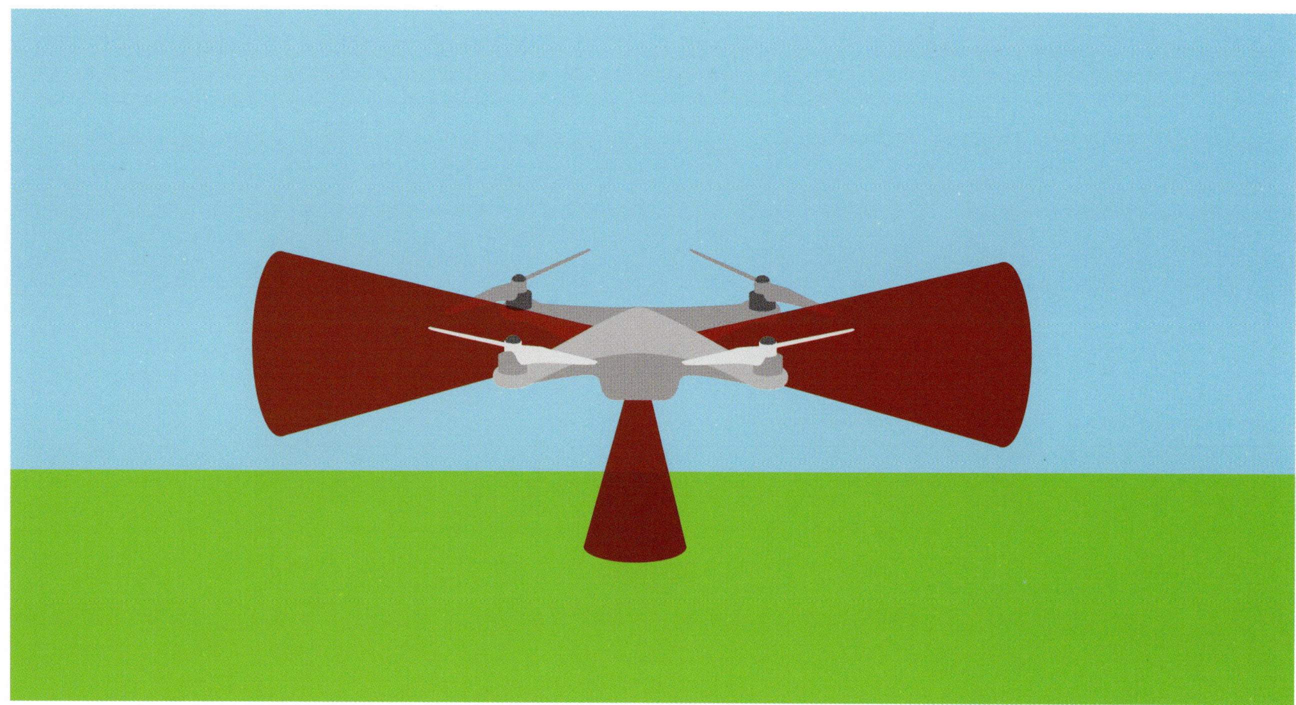

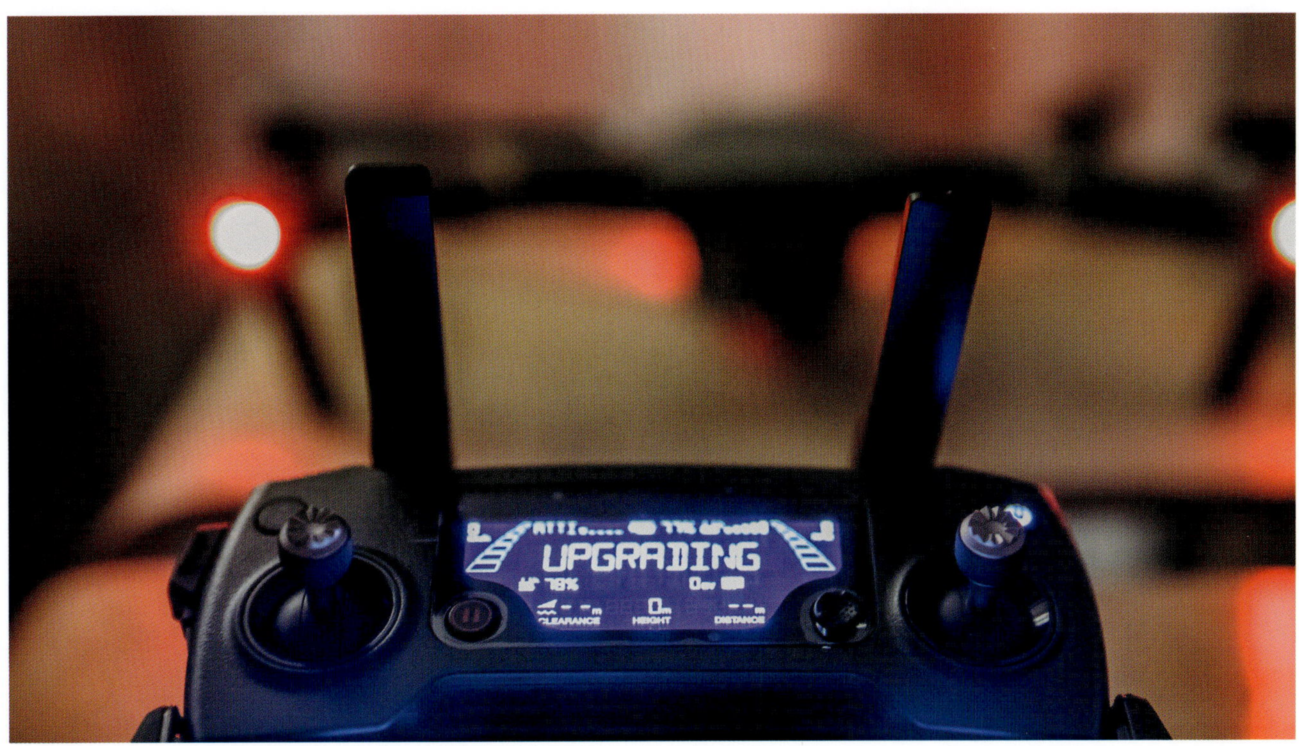

FIRMWARE UPDATES & APP VERSIONS

In the same way that software and apps are continuously updated, the embedded firmware in your drone—and its remote transmitter—will likely be updated periodically by the manufacturer. This could be to add extra features or to eliminate bugs and improve functionality. It is likely that your app will notify you when new firmware is available and that you should update your drone and transmitter.

Despite the availability of new firmware, the drone should keep operating as normal with the older firmware. In fact, unless you have a specific problem (or the manufacturer has warned that it's a critical update), I often find that it's wise to wait a week or two and let other people check that the new firmware works as it should. It is not unheard of for firmware updates to introduce new problems, so letting a few other pilots test it first is not a bad idea!

Once you have decided to update, it's usually a matter of downloading a file from the manufacturer's website, transferring it to a compatible memory card, and then powering up the drone (indoors, without propellers, making sure you have a full battery). It should detect the new firmware file and go through an updating procedure—this may take up to 20 minutes depending on the size of the update. The remote transmitter firmware can often be updated through the app itself. The exact procedure will vary from model to model, but clear tutorials are available online.

▲ After updating the firmware on your drone and transmitter it's a good idea to ensure you have the current version of the app as well.

FLYING SKILLS

Much like any motor skill, flying a drone is something that develops with practice. So, before you start worrying about camera settings, composition, and any of the finer points of photography (which we'll look at in the following chapter) you need to be able to fly your drone safely and instinctively.

For stills photography, the ability to get the drone to where you want it and back in a safe manner is really all you need to be able to do. However, if you want to shoot good video, the way you fly your drone is very important—you need to be able to fly precisely and smoothly.

Taking to the air for the first time with your brand new drone can be intimidating, but it needn't be a harrowing experience. Flying a modern, GPS-stabilized drone safely is a relatively easy task, and one that is well within the reach of most people. While advanced flying skills can take years of practice to perfect, you can be up and flying safely in very little time, providing you follow a few easy guidelines.

If you're nervous about trashing your expensive camera drone on its maiden flight, a good strategy is to buy an indoor microdrone to practice with. These can be bought pretty cheaply, have the same control layout as their larger cousins, and are a lot of fun.

As microdrones are not GPS stabilized (GPS doesn't work well indoors anyway), they will tend to drift after take off if you leave them in a hover without touching the controls. Some models allow you to adjust the trim on each axis of movement until a satisfactory hover is achieved, or you can just use the controls to keep it steady. At the very start it is sensible to keep the drone facing the same way at all times (nose pointing away from you), so you shouldn't need to touch the rudder control.

▼ Using a microdrone is a great way of practicing your flying skills. They're not as sophisticated as their larger brethren, but the controls are often very similar, and they're far less expensive to replace if things go wrong!

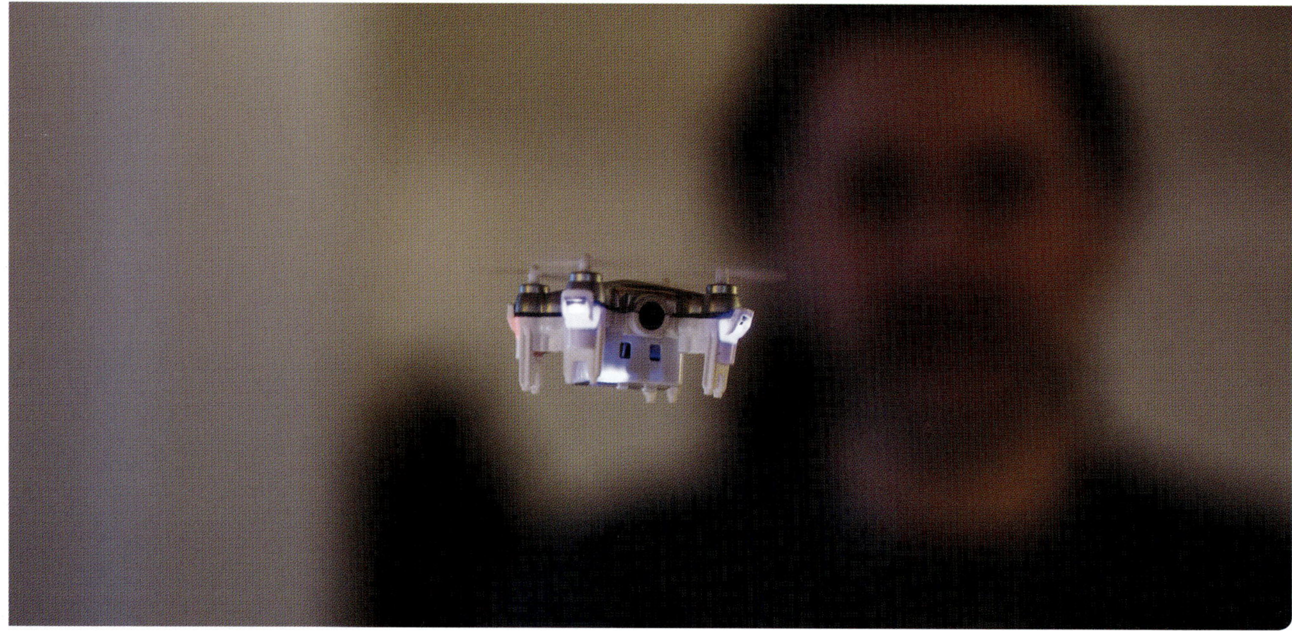

Up & Down (Throttle)
The up/down movement on the left stick increases or decreases throttle, making the drone move up or down in altitude. Typically the drone will hover with the stick centered (throttle at 50%). On many drones, holding the throttle at its lowest position for more than a few seconds will kill the motors—this is good after you've landed, but not so good mid-flight.

Up & Down (Elevators)
The up/down movement of the right stick controls the forward/backward movement of the aircraft. This is achieved by simply tilting the aircraft forward: the farther forward you push the stick, the faster the aircraft will move. Note that the direction is relative to the heading of the aircraft.

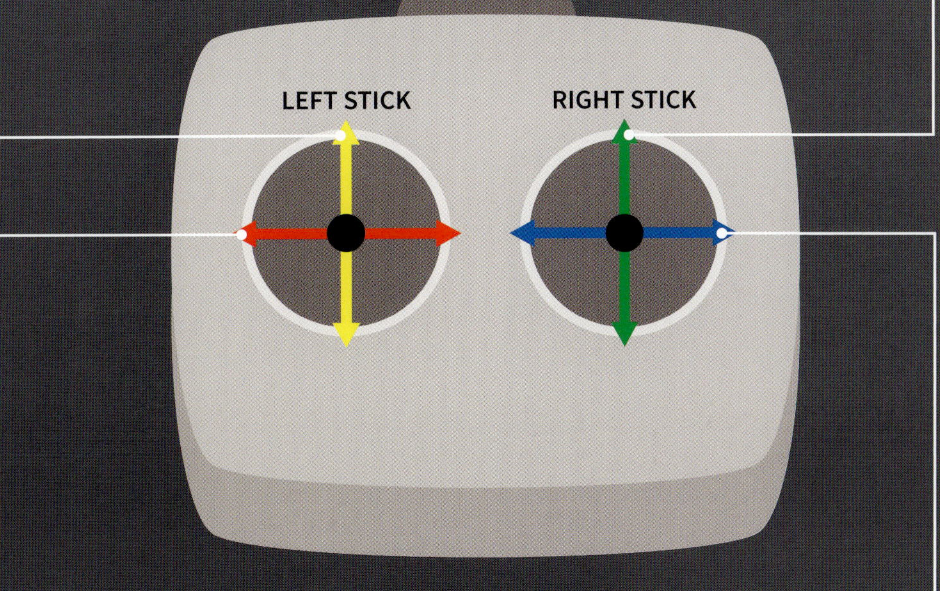

Left & Right (Rudder)
A left/right movement on the left stick will "yaw" the aircraft (rotate it about its own axis): right is clockwise, left is counterclockwise. It's wise to minimize your use of the rudder to start with, as all of the controls are relative to the direction of the aircraft, not relative to you. So, if the drone is facing you and you put the aileron stick to the left, the aircraft will move to ***its*** left, which will be ***your*** right.

Left & Right (Aileron)
The left/right movement of the right stick controls the movement of the aircraft to the left or right relative to its own heading. This is achieved by increasing the relative speed of the two rotors on the left or right of the aircraft.

The most common control layout of a transmitter is known as *Mode 2*, as shown here. This is usually the default configuration and unless you have good reason it should not be changed. This illustration shows the flying controls, but in addition there will be various dials and switches for flight modes and camera/gimbal control, depending on your drone and controller.

PUTTING IT ALL TOGETHER

As a beginner you will probably spend a while using just one or two axes of control at a time, leading to a boxy, angular flight path. As your confidence increases, you will start to use all the axes simultaneously, flying in smooth, graceful arcs.

Whether you choose to practice with a microdrone indoors, or go straight to a larger, GPS-stabilized drone outdoors, there are a number of exercises you can use to gradually improve your flying skills. Just flying about randomly might help with your confidence, but if you want to improve your precision, it's a good idea to have some goals.

HAND CATCHING

Situations may arise where it is easier or safer to "hand-catch" the drone, rather than landing in a tricky spot. This should only be attempted by experienced operators with smaller drones, as there is the potential for serious injuries from the spinning propellers. Be aware that if the aircraft has obstacle avoidance or altitude sensors, it may try to avoid being hand-caught. It should be obvious, but if you ever try to hand-catch your drone, the main aim is to keep your fingers as far away from the moving propellers as possible.

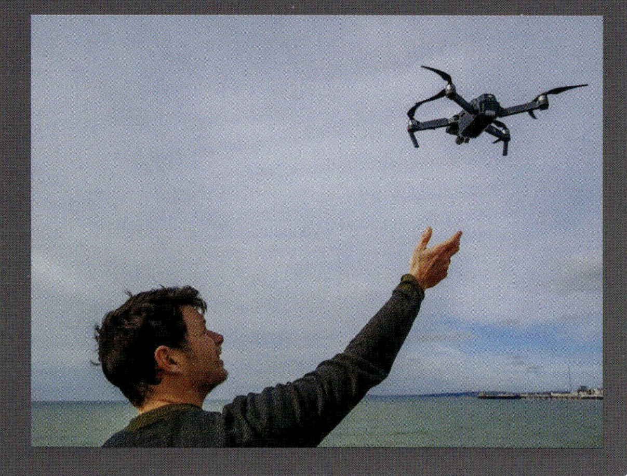

If you are practicing outside, always ensure you have plenty of space, are a good distance from trees or other obstructions, and there are not many other people close by. Remember you should not be flying close to the general public, or directly over them at any height. Many drones have a beginner mode that limits the drone's maximum speed, maximum height, and distance from you—it might be worth putting it in this mode for your first few flights.

Taking off and landing are perhaps the trickiest parts of drone operation. The take-off and landing spot should be clear of obstacles in all directions and be smooth and flat—even relatively minor unevenness in the ground can encourage a drone to tip over, possibly damaging the propellers. But there is nothing to be worried about—just make sure you run through the items on the preflight checklist on page 69 before take-off.

Take off

With most GPS-enabled outdoor drones you have to "unlock" the motors before you can up the throttle to take off. This is usually achieved via the CSC (Combined Stick Command). Once the motors have started spinning, confidently and smoothly return both sticks to the center position and continue pushing up on the left (throttle) stick until the drone is well clear of the ground (say 2 yards/meters). Avoid hovering at 50% throttle very close to the ground, as the drone might touch and tip over.

Landing

When you come in to land your drone, you should make it a positive move. Avoid drifting around very close to the ground, as you risk tipping the drone on contact. I tend to approach to a height of about 1 yard/meter from the ground; descend vertically to around 4in (10cm); and then bring the throttle hard down, which should drop the drone down vertically. Once it's on the ground, cut the motors entirely. After the motors have stopped you should immediately approach the drone and power it down.

▶ Only fly your drone out over the sea when you feel you have a good grasp of both the drone's operation and the weather conditions. An unexpected Auto-Landing could be disastrous!

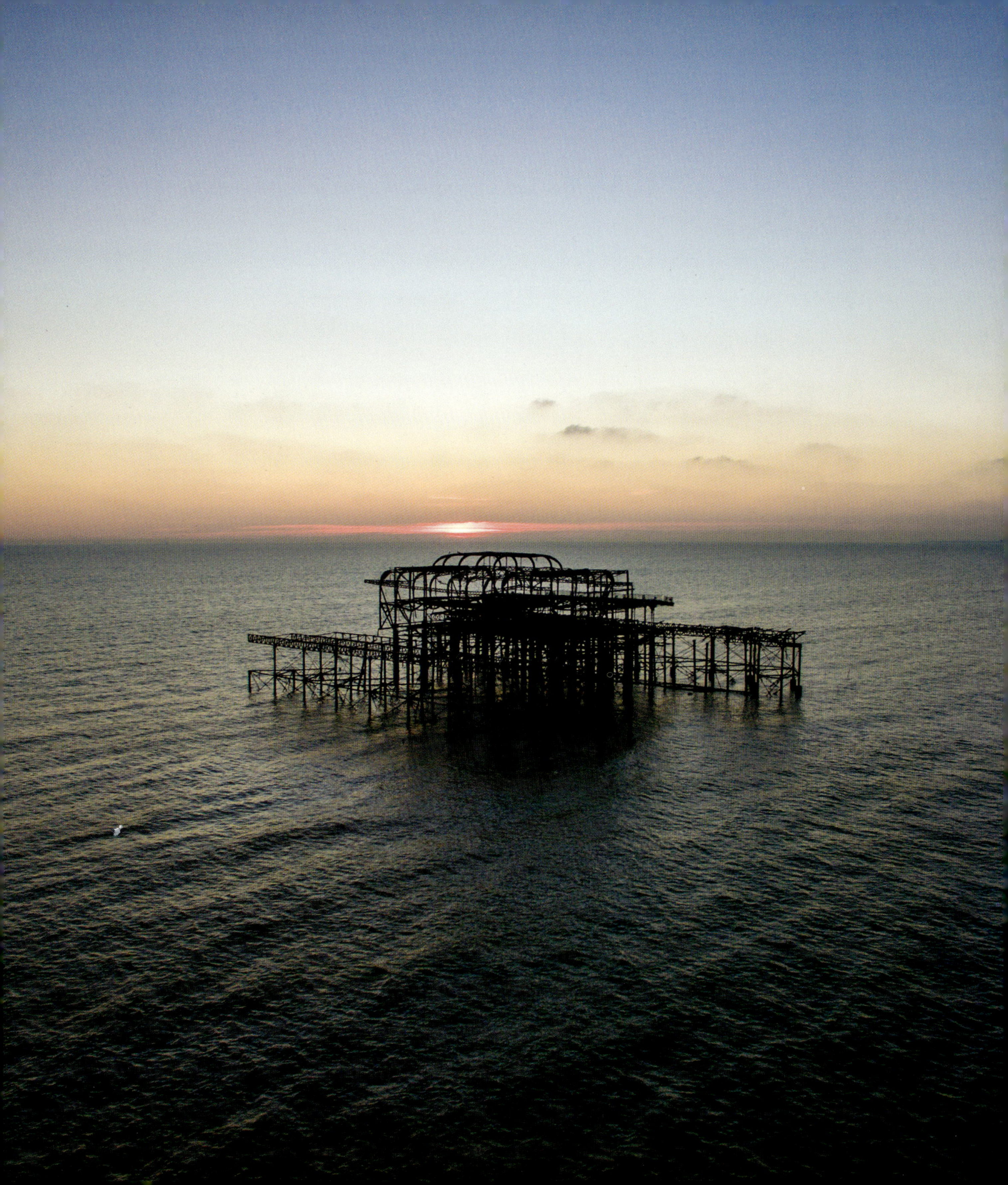

PROFILE: FREDRIK CHRISTIANSEN

Fredrik Christiansen is a postdoctoral research fellow at the Cetacean Research Unit at Murdoch University, Western Australia, whose work involves identifying and measuring the body condition of individual whales.

As soon as relatively inexpensive and reliable drone technology became available, Fredrik realized that he had the "perfect" tool for his work. Highly portable, cheap to run, and quick to deploy, these aircraft could put a high-quality camera directly over a whale group within minutes of spotting them, in contrast to the expense and difficulty involved with conventional manned aircraft, or the limited viewpoint of a boat.

"Drone technology has made it possible to measure the size and body condition of free-living whales without disturbing them. I did have an initial concern over whether the noise from the drone could upset the whales, so I investigated how the noise penetrated into the water column. I did this by flying the drone over an array of underwater microphones and measuring the noise level of the drone at different altitudes. I found that when flying above 30ft (10m), the noise level underwater—even at a depth of just 3ft (1m)—was below the ambient noise levels."

However, it's a different story for terrestrial animals and birds, where there is a real concern about drones affecting wildlife. According to Fredrik "it has been shown that various species of birds avoid or sometimes even attack drones, while another study showed that the heart beat of bears increased substantially in the presence of drones, which is a sign of stress."

▶ **Fredrik photographed this pair of southern right whales— a mother and calf— off the coast of Western Australia using a DJI Inspire 1 Pro.**

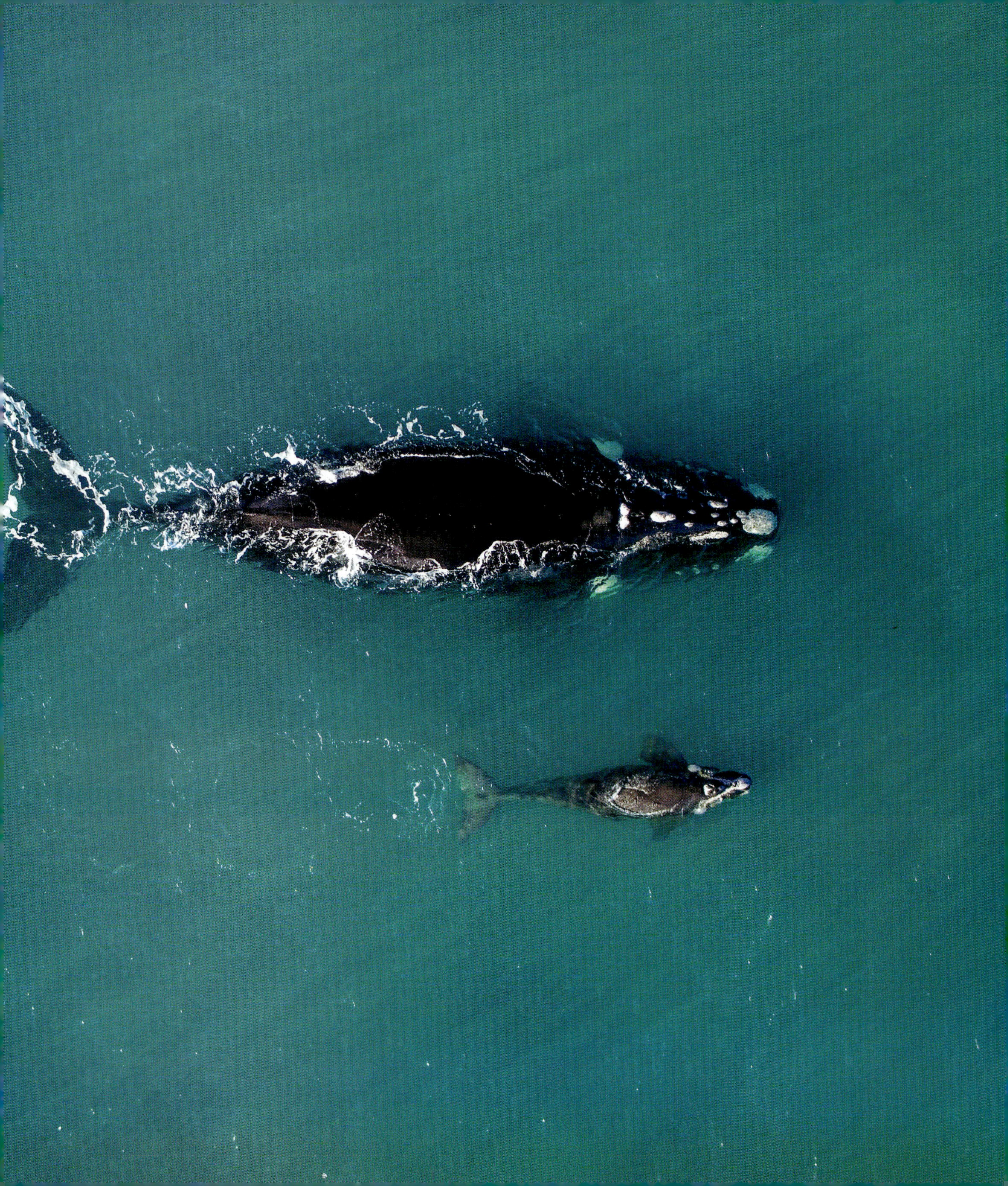

FLIGHT MODES

Most drone manufactures offer a variety of flight modes that can make flying your drone much easier, but the choices can be confusing at times. The following explanation is based on the DJI system, as implemented by its newer drones, but many manufacturers have their own versions of these systems (albeit with different names).

The first and most important flight mode for the novice is Beginner Mode, which restricts the maximum speed of the drone. It also limits the distance it can travel from the pilot to approximately 90ft (30m).

Additional "basic" flight modes are GPS, OPTI (Optical), and ATTI (Attitude). GPS mode is used outdoors when the drone can pick up enough satellites for an accurate position lock. This will make the drone more stable and less liable to drifting with the wind, although if you fly under trees or close to buildings you could lose the GPS signal.

OPTI (or Optical) mode is typically used when GPS is unavailable. In this mode, downward-facing optical sensors are used to stabilize the drone. However, for this to work the drone needs to be no more than 10ft (3m) from the ground, flying over a suitable surface, and have sufficient light for the sensors to do their job.

ATTI (or Attitude) mode is employed when neither the GPS nor the optical sensors can be used. The drone will hold its "attitude" (return to a flat hover) when there is no stick input, but it will be subject to being blown by the wind. If you are traveling in a particular direction, you will have to actively brake by applying some thrust in the opposite direction.

In addition to these basic modes, many newer drones have a range of "intelligent" flight options, as summarized on the following pages.

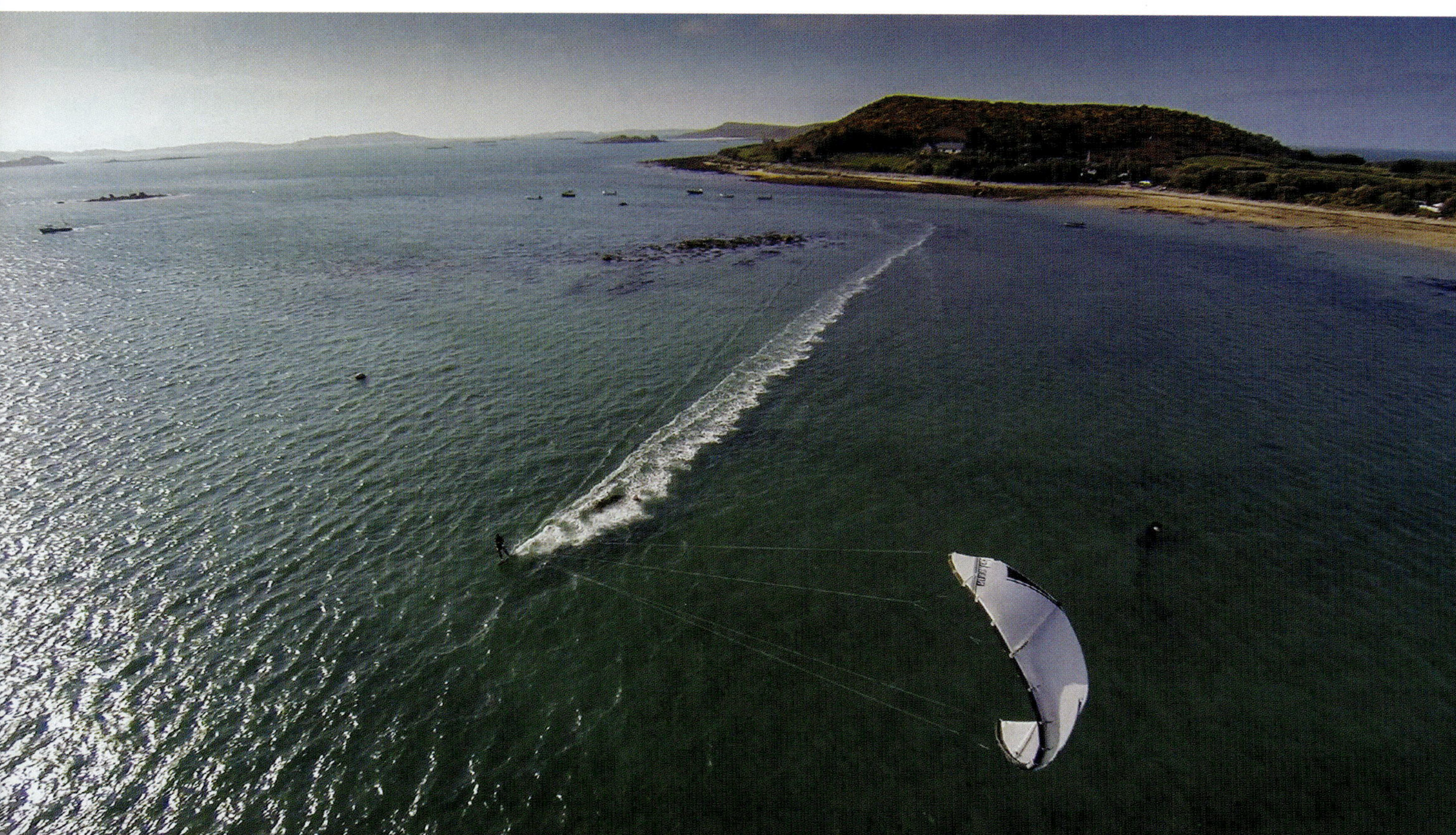

BASIC FLIGHT MODES

MODE	WHEN TO USE IT	PROS	CONS
Beginner	Use while learning.	Your drone can't go too fast or fly too far away.	Your speed and range are both limited.
GPS	Use outdoors in clear open areas. This is the mode used most frequently.	Makes the drone easier to fly, particularly in windy conditions. Adds stability and gives accurate RTH ability.	Only works outdoors and with a good view of the sky. GPS signal can drop out with little warning.
OPTI (Optical Sensors)	Use when flying indoors or under tree cover (only with good light).	Adds stability when GPS is unavailable.	Only effective up to an altitude of around 10ft (3m) and with good light.
ATTI (Attitude)	Use when neither GPS nor OPTI are options.	Great for practicing drone control. Many pilots find it easier to control steady speed increases in this mode (in video tracking shots, for example).	Drone will be affected by the wind and it takes some practice to master.
Sports	Use when you need to fly as fast as possible!	Greater top speed and maneuverability. Uses GPS when available.	Collision sensors disabled.

◀ A kitesurfer at speed. Sports mode not only lets you fly quicker to keep up with fast-moving subjects, but also enhances your ability to combat strong winds.

INTELLIGENT FLIGHT MODES			
MODE	WHEN TO USE IT	PROS	CONS
Course Lock (Headless Mode)	Use when you want to fly in a straight line, whilst yawing the aircraft.	Some video shots are easier to achieve in this mode—for example, when you want to follow someone or something moving in a straight line, while yawing the drone.	The right stick directions are relative to the initial heading of the aircraft at take off, not to its current heading.
Home Lock	Use when you want the stick inputs to be relative to the home point.	If the drone is distant and you are unsure which way it is facing, pulling down on the right stick will always bring it back toward you.	Don't get confused over heading direction!
TapFly	Use when you want to tap on the screen to make the drone fly in a certain direction (tapping a new point will make it change direction).	Easy to use and you can still use the sticks to control the aircraft.	Need to remain vigilant for obstacles, particularly small ones that may not be detected by any obstacle avoidance.
Point Of Interest	Use when you want the camera to remain aimed at a specific subject or point.	Takes the hard work out of getting orbit shots.	Beware of sideways collisions with obstacles, particularly if your drone only has forward-facing sensors.
Follow Me	Used to make the drone follow whoever is holding the controller (or uses the GPS position of a phone/tablet).	Great if you want the drone to follow you while doing sports or activities.	The subject needs to have the controller/device. Need to watch out for collisions with obstacles.

INTELLIGENT FLIGHT MODES			
MODE	WHEN TO USE IT	PROS	CONS
Active Track	Used to make the drone follow a specified person or object by tapping on the screen on the app. Further sub-options include "profile" (for a tracking shot from the side); "spotlight" (keeps the camera aimed at the subject, regardless of how you fly the drone); and "trace" (the drone will automatically orbit a moving subject).	Takes the hard work out of tracking a moving subject. Allows a variety of options.	Only use when the subject stands out clearly against the background and beware of rapidly changing direction or going behind obstacles. Need to be aware of sideways collisions with obstacles, particularly if your drone only has forward-facing sensors.
Gesture Mode	Used when you want to take selfies and instruct the drone to track you by making gestures at it.	Can be fun, but a "gimmicky" feature for most users.	Need to be aware of obstacles and always keep the controller to hand in case of unexpected drone behavior.
Waypoints	Use when you want the drone to fly a preplanned course of waypoints. This can be useful for repeating a flight plan many times over a long period (for creating a timelapse sequence, for example).	You can choose whether you want to be able to change the way the drone faces during the automatic flight, or use a pre-planned camera direction.	Need to be vigilant of hazards and obstacles.
Terrain Follow	Use when you want the drone to automatically maintain the same height above the ground when flying over a gentle slope.	You don't need to worry about hitting the ground when flying up a gentle incline.	Will not work if you have a sudden steep section like a cliff. It's only effective at altitudes of around 1–33ft (0.3–10m). Does not work for descending flights.

GETTING TO GRIPS WITH FLIGHT

Although the simple act of flying will improve your skills, setting yourself specific flying tasks will really help develop your control and precision. The following are some basic exercises to get you started, but as you improve it's really worth continuing to set yourself challenges, as the more control you have over your drone, the safer you will be.

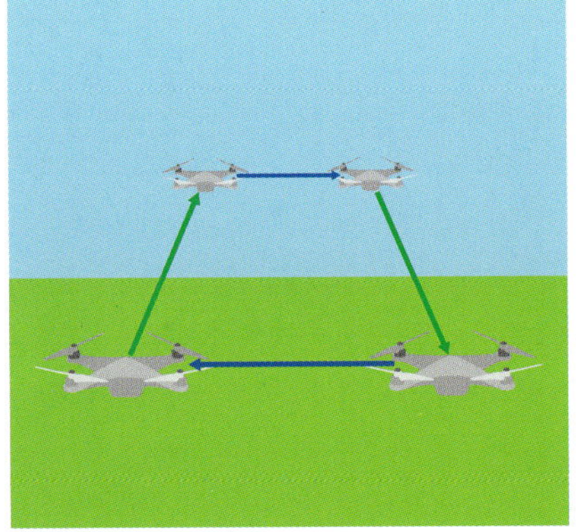

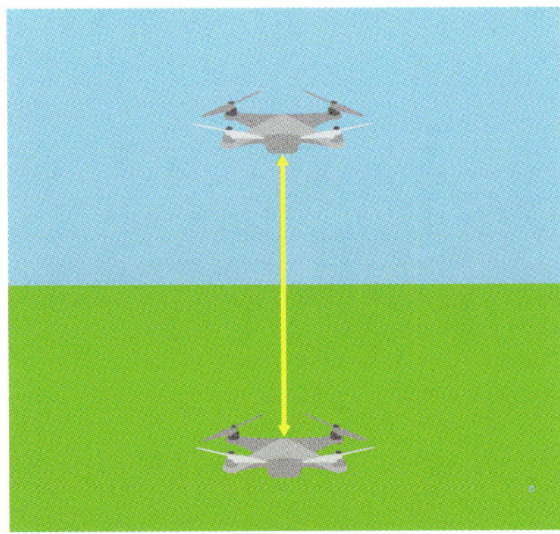

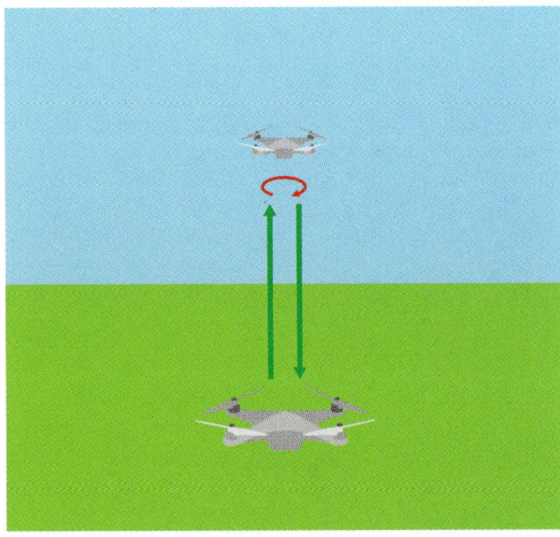

DRAW A SQUARE

A useful early exercise is to fly your drone in a square pattern, using only the aileron and elevator controls. If your drone has GPS lock and a decent altitude lock this should be fairly easy.

VERTICAL RISE

Another straightforward exercise is to fly straight up and down, using only the throttle control. After you've practiced your vertical ascents and descents you might want to try diagonals, using some right stick in the desired direction in addition to the throttle control.

BACK TO FRONT

It takes some practice to remember that the controls are all relative to the drone, not to you, so fly away from yourself, stop in a hover at a distance where you can still clearly see the drone (say 20 yards/meters), and then slowly rotate the drone with the rudder control until it is facing directly toward you.

You will now find that pushing the right stick forward will bring the drone back toward you (this is because you are moving forward relative to the drone). Pushing the right stick to the left will move the drone to its left, but *your* right. Some people pick this up quicker than others, but it's just a matter of practice. The key is to be aware of the direction your drone is facing.

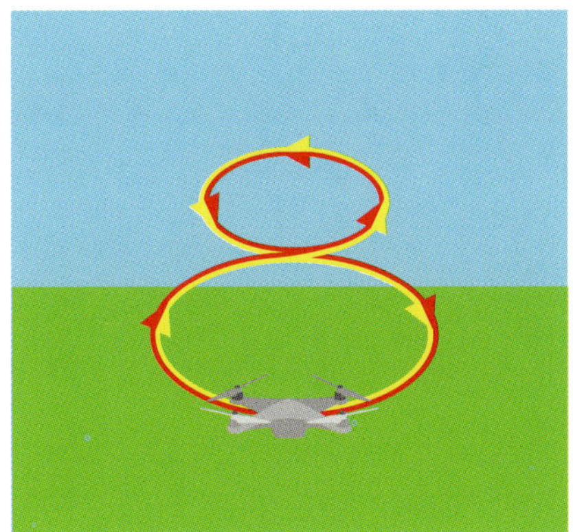

FIGURE OF 8

When you feel you have a good feel for the throttle and right stick controls, start experimenting with steering using the rudder. A good exercise for this is to practice flying a "figure of 8." This is achieved by combining the right stick forward (elevator) with the left stick side to side (rudder). Remember that the controls are reversed when the aircraft is heading toward you.

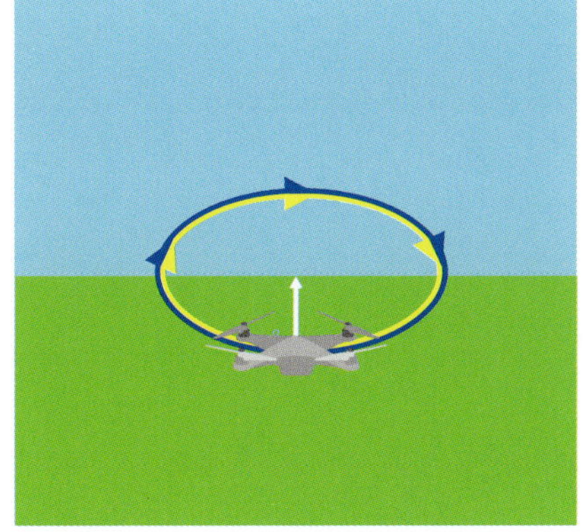

PERFECT CIRCLES

Another good flight practice is to try flying your drone in a perfect circle. There are several ways to achieve this, but perhaps the simplest is to use only the right side controls.

From the videographer's point of view, a particularly useful skill is to fly a perfect circle round an object, while keeping the camera aimed directly at it. This is achieved by pushing the aileron control to the right, while the rudder control is pushed left (or vice-versa). It takes practice to master and the rudder movement will generally need to be much smaller than the aileron movement.

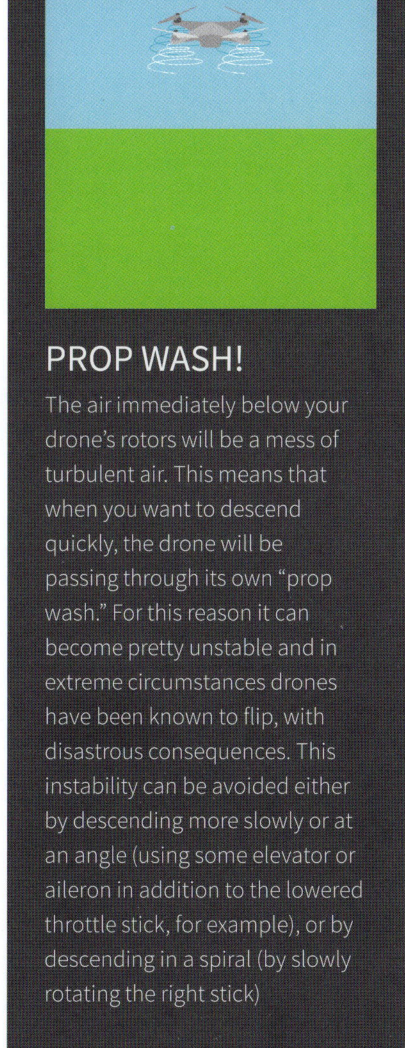

PROP WASH!

The air immediately below your drone's rotors will be a mess of turbulent air. This means that when you want to descend quickly, the drone will be passing through its own "prop wash." For this reason it can become pretty unstable and in extreme circumstances drones have been known to flip, with disastrous consequences. This instability can be avoided either by descending more slowly or at an angle (using some elevator or aileron in addition to the lowered throttle stick, for example), or by descending in a spiral (by slowly rotating the right stick)

LINE OF SIGHT

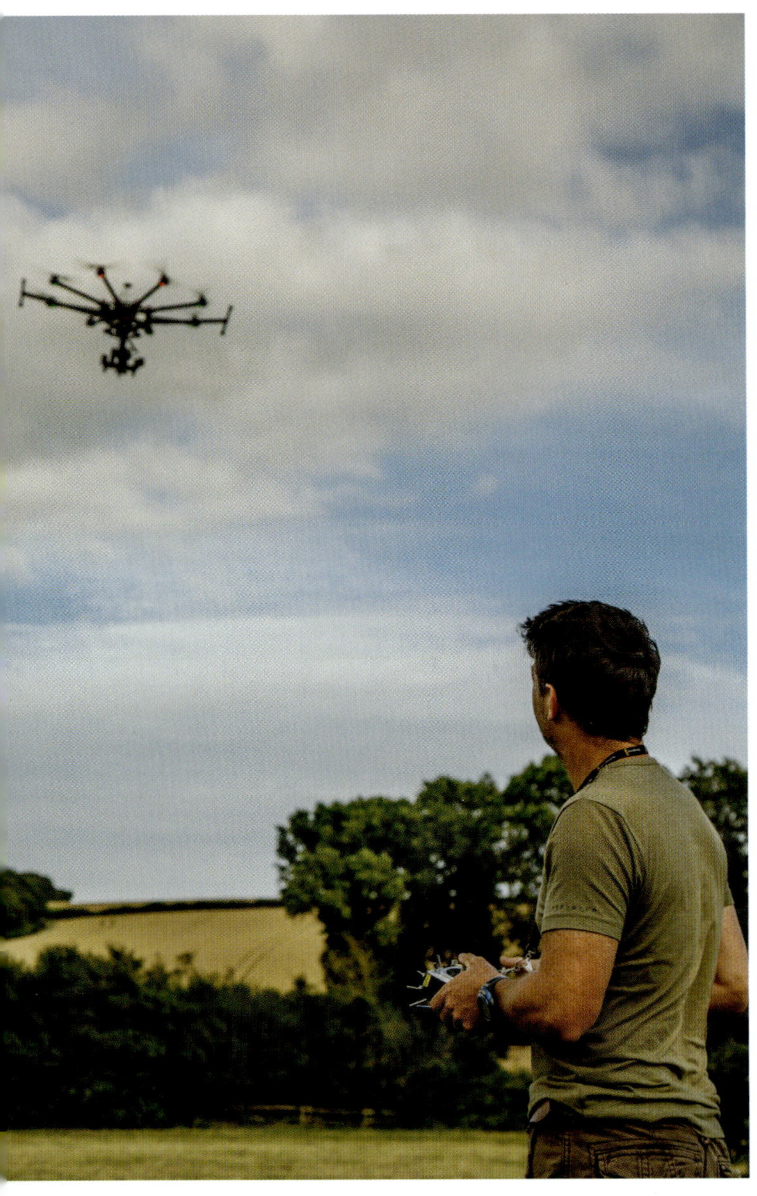

▲ To minimize the risk of collision you should always maintain clear line of sight with your drone.

It is always tempting to fly your drone miles away from you, relying on the image on your monitor and your flight data to tell you what your drone is doing. For a number of reasons this is not a good idea:

- In many countries it is illegal to operate a drone beyond visual line of site.
- At large distances you will lose control frequency contact with your drone.
- You might lose control of your drone if it flies behind some sort of obstruction.
- You are likely to lose the video feed as distance increases.
- It's harder to avoid collisions with obstructions if you're relying solely on the video feed.

My general philosophy is that I want to be as close to the drone as I possibly can be, without compromising the shot I need to get. This is largely because depth perception is much better at shorter distances. Once your drone is 50 yards/meters (or more) away from you, it becomes quite hard to tell how close it is to obstructions. Obviously this is not as much of an issue in a wide open and completely flat area, as you can see there are no obstructions, but I have heard of plenty of instances of pilots flying their drone fast and low, unaware of a gentle slope in the ground until they have ploughed straight into it. Fortunately, this type of crash is becoming less common as safety features such as acoustic altimeters become more common, but it still pays to be aware of subtle topographical features, as well as more obvious obstacles!

FPV

FPV, or First Person View, refers to a method of flying a drone where the pilot relies solely on what he sees through the video feed from the drone's camera. The most immersive way of doing this is using FPV goggles, which range from relatively expensive, dedicated HD goggles containing two small video monitors to inexpensive enclosures for a smartphone.

FPV aficionados typically race small and very fast drones through purpose-built racecourses, negotiating tricky obstacles at breakneck speed. This is a great way of improving your flying skills, but it's not for the faint-hearted!

For drone photography purposes it's usually better for the pilot to be able to observe his drone directly, although if you have a separate camera operator, he or she will be much more interested in the video feed and may even prefer to use FPV goggles to "see" what the camera's seeing. There are even setups where the operator can pan and tilt the camera by moving their goggle-equipped head with a virtual-reality type setup (using the accelerometers in the smartphone).

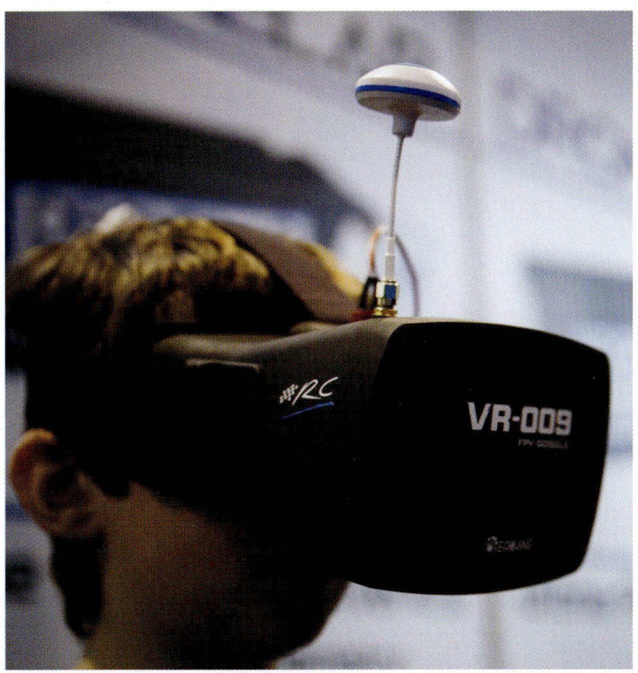

▲ FPV goggles offer a truly immersive flying experience, but with no peripheral vision, you should have an observer on hand to warn of unanticipated hazards.

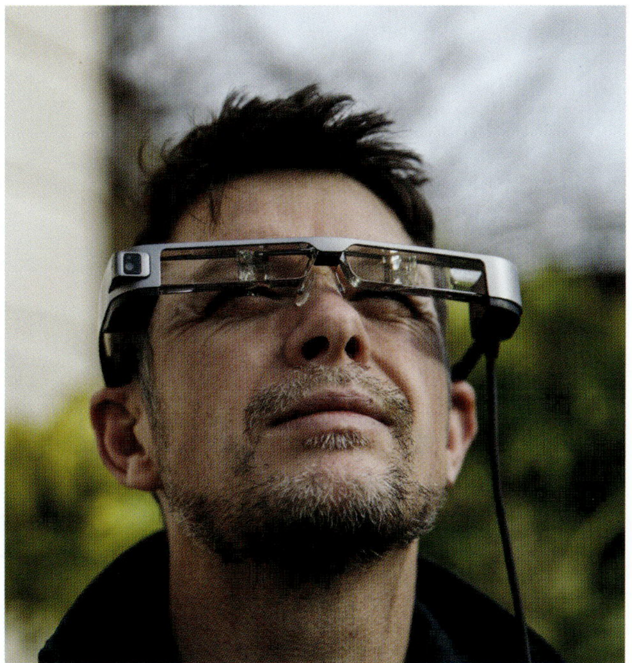

▲ Augmented reality glasses give you a "head-up display" (HUD) of your drone's camera view and flight data, but also allow you to maintain direct line of sight.

PROFILE: FELIX VON AULOCK

Volcanologist Felix von Aulock is a postdoctoral research associate at the University of Liverpool, UK. In 2016 he traveled to Guatemala, Central America with a drone to get an aerial view of the erupting Santiaguito volcano.

"Drones allow us to get a close view of active volcanoes while maintaining a safe distance. We use this data to monitor the activity of the volcano, and measure changes in the crater between and during explosions. Work in volcanic environments is often very taxing (both on the gear that is constantly covered in volcanic ash and on volcanologists), so we appreciate the recent development of smaller, lighter, and more reliable drones. For us, getting the data is paramount on our expeditions, so capturing the important data, reliable backups, and video downlinks has priority over the visual esthetic of the footage."

Felix's unique aerial footage of an eruption in progress made spectacular viewing, which was featured on the National Geographic television channel. More than that, Felix could also use it to help construct computer models that scientists can then use to predict the effects of eruptions and safely manage the dangers presented by active volcanoes.

▼ Santiaguito volcano erupts! For this video sequence, Felix used a DJI Phantom 2 with a GoPro Hero 3 camera attached, shooting 2.7K video at 30 frames per second. The main challenge was flying close enough to get spectacular footage, but not so close that it endangered the drone.

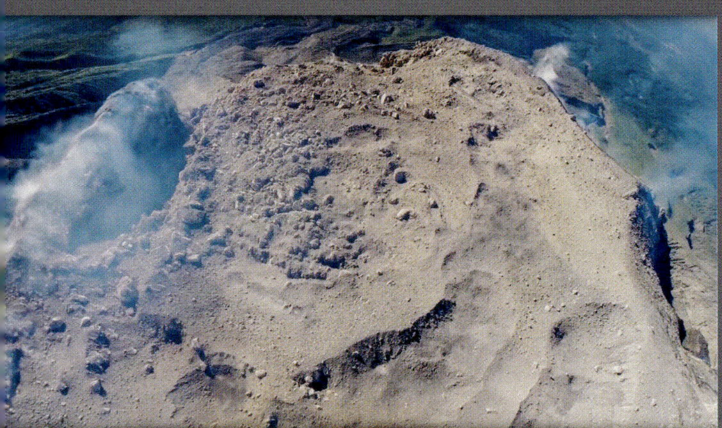

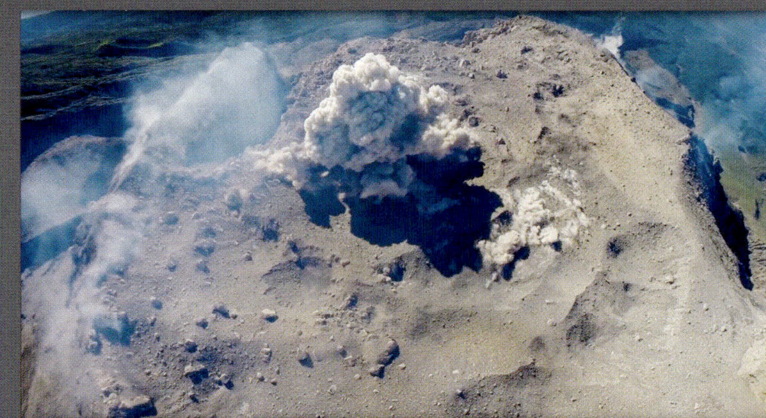

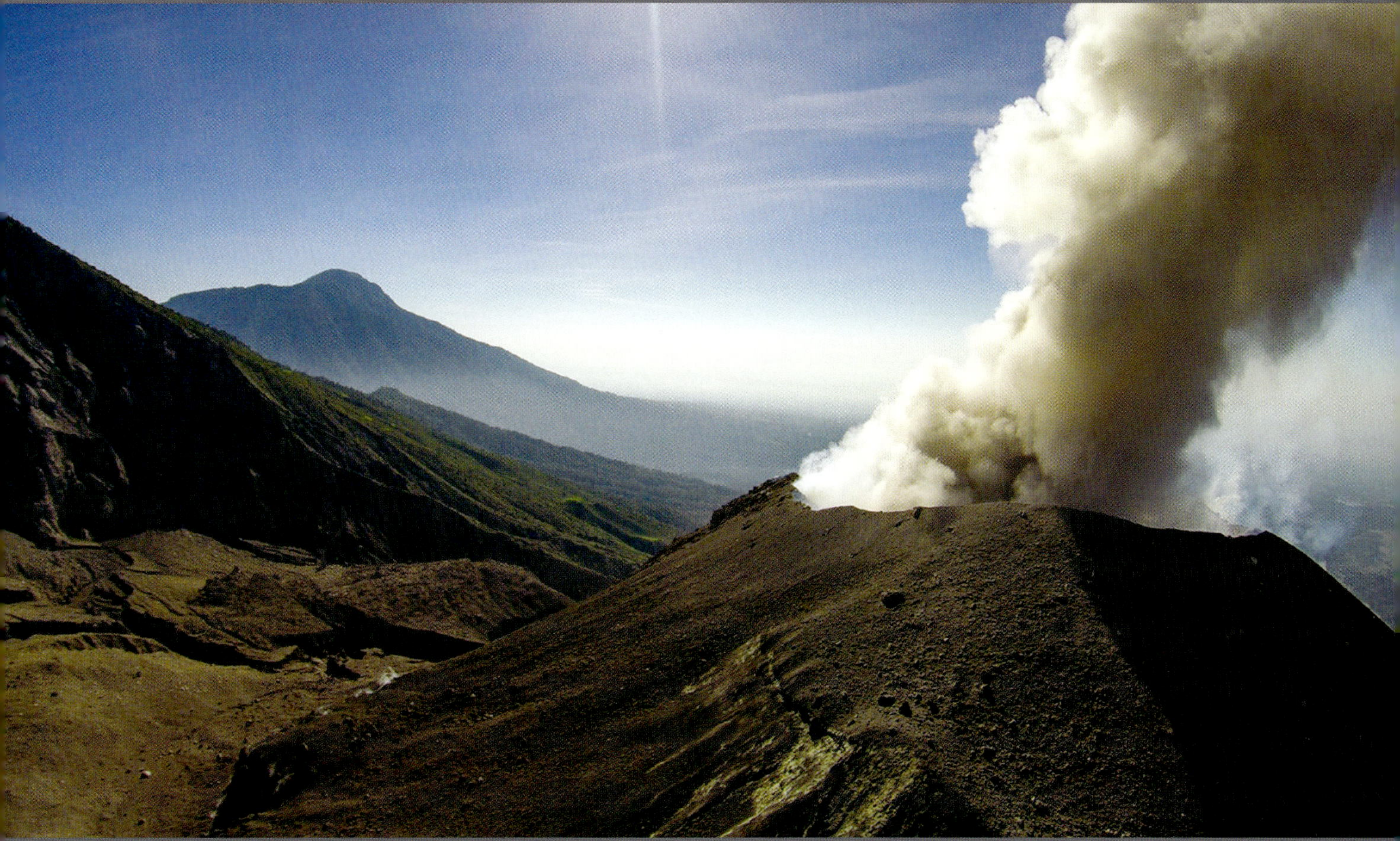

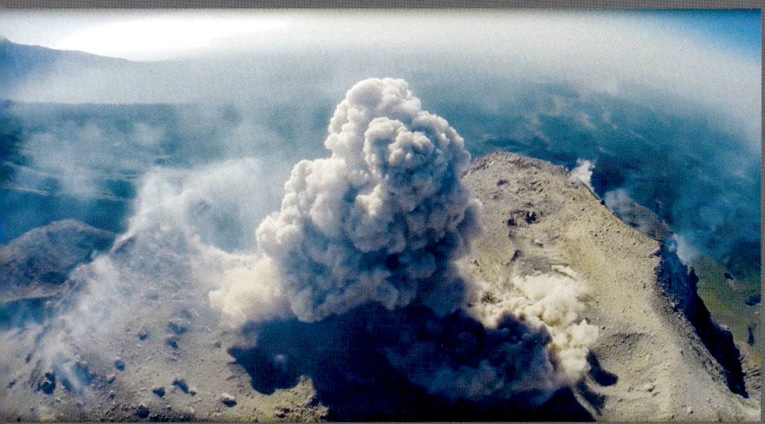

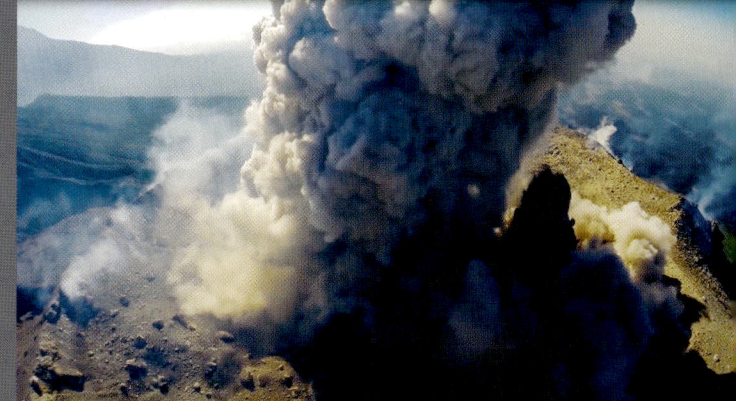

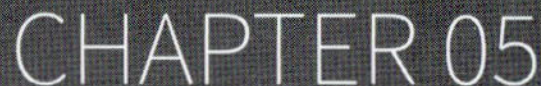

CHAPTER 05

STILL PHOTOGRAPHY

With the initial excitement of your first serious outdoor drone, it's easy to imagine that you'll just fly this wonderful piece of kit into the air and start taking epic shots. The truth, however, is that even from 400ft (120m) in the air it's still possible to come home with mediocre results!

In this chapter you will see how to approach still photography with a drone from a creative point of view, as well as the technical aspects that you need to be aware of. I will also talk about some of the more unusual and specialized techniques you can use for your aerial photographs, and suggest how best to process your digital shots. However, this can only be achieved once flying your drone has become second nature, so be sure to put in the practice!

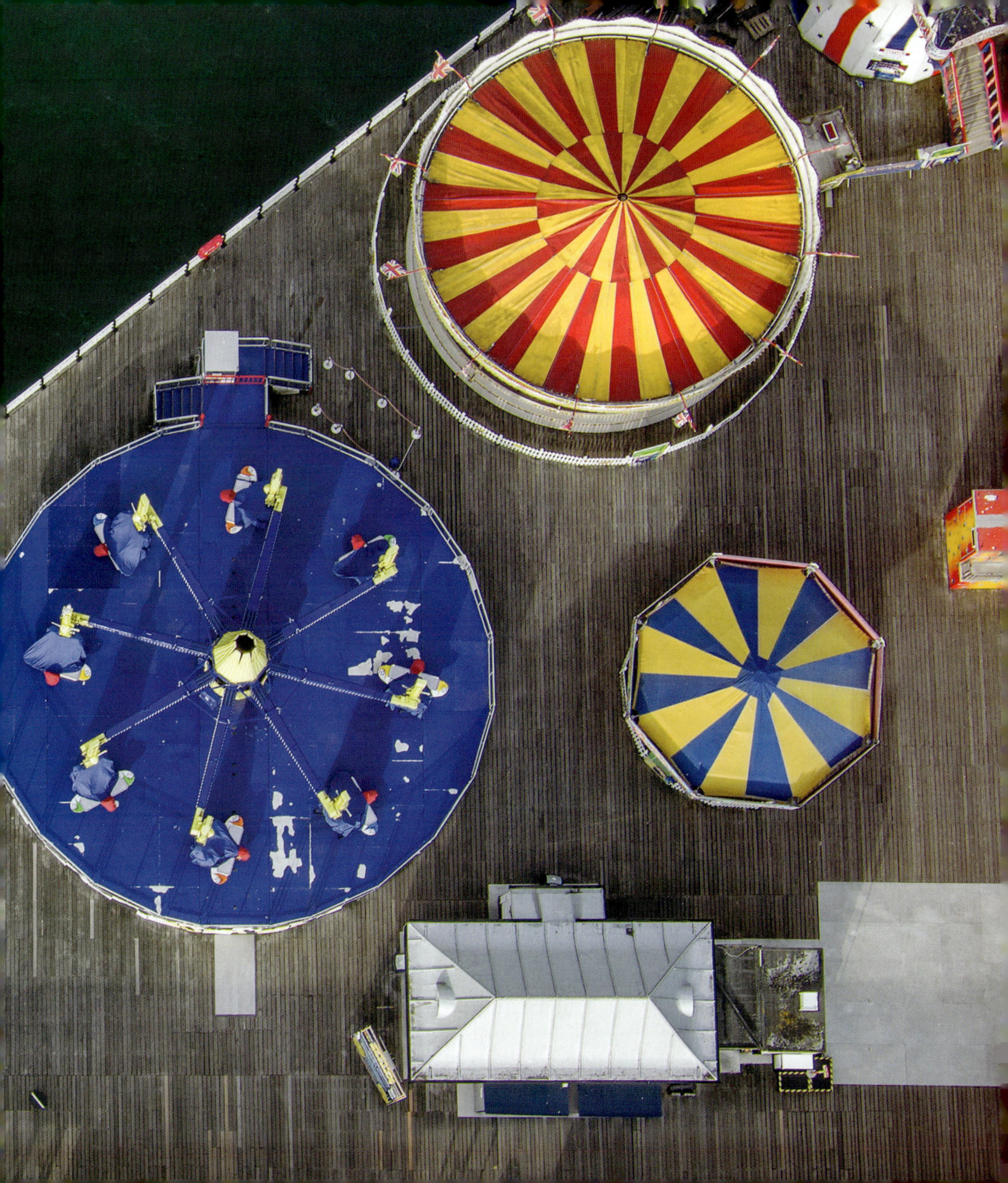

WHAT MAKES A GREAT PHOTOGRAPHER?

The attributes that follow are those that I believe all great photographers share, regardless of whether their cameras fly or not. So while this is not unique to aerial photography, it is worth considering at this point, as it is arguably a more important factor in achieving great photographs than any equipment-led considerations.

The fact is, very few truly outstanding photographs are the result of chance. They come about through a process. They have been imagined, carefully planned, often rethought and reshot (possibly on multiple occasions), sifted out from multiple variants, and perhaps lovingly fine tuned in post-production.

Of course, photography is also a highly subjective art form, so an image that wows one person may leave the next cold. With that in mind, perhaps the most important thing—particularly for the enthusiast photographer—is the enjoyment to be had from a feeling of improvement and progression.

CREATIVITY

This is an overused term, but the first step to taking a great photograph is imagining it. In the case of aerial photography you need to be able to imagine what the world might look like through the eyes of a bird, or from an airplane. I don't believe this is something you either have or don't have from birth—it comes with experience and practice.

As you fly, you will often see unexpected patterns or compositions on your drone monitor during a flight. However, it might be that the light isn't quite right on that day at that time, or maybe you suddenly realize you could get a similar but better shot elsewhere. The key thing here is to not only recognize a potential photograph, but to be prepared to return on a different day, at a different time, or to relocate to get the best shot.

OBSERVATION, ANALYSIS & CRITIQUE

It should go without saying that the ability to use your eyes and brain is fundamental to good photography. An excellent photograph should grab you on a gut level—you love it, but you may not know why. For some photographers (maybe not all), analyzing what you like about the photograph in terms of its composition may be revealing. Equally, when you have taken a photograph that doesn't quite work, ask yourself why that is. With practice, this process becomes much easier and quicker, and you will soon find you can quickly identify what you feel works well and what doesn't.

PERSISTENCE

Once you have critically assessed your photographs and pinpointed areas that need improvement, the necessary drive and persistence to go back and take a photograph again, and again with slight changes in composition or lighting can be the difference between a good shot and an outstanding one.

I have frequently returned to the same spot for both ground-based and aerial shots to try and rectify a flaw in a previous version: the sun was not rising at quite the right angle (different season); the tide was wrong for a coastal shot (different time of day or tidal cycle); there was too much fog or too little fog; too much wind; too many people; wildlife not cooperating—the list of potential factors is almost infinite. Many of these shots still exist only in my imagination, but I live in hope that one day I'll nail them!

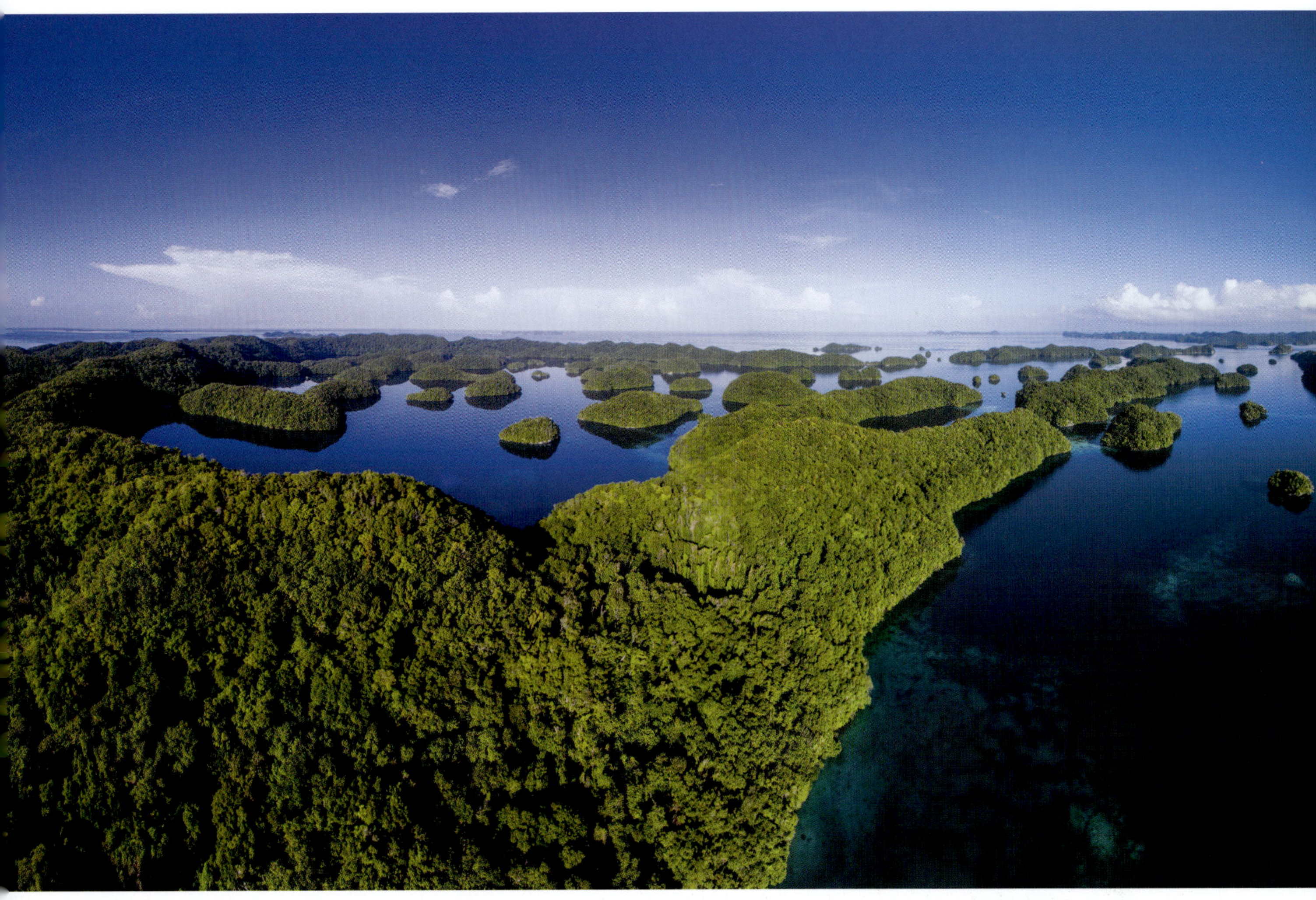

TECHNICAL ABILITY

I see technical ability as a toolbox that enables you to correct some of the shortcomings you might find in your own photographs. It takes little technical ability to set the camera to automatic, but if the results are unsatisfactory from a technical point of view, it's great to understand what went wrong and know how you can do better next time. Was the photograph under- or overexposed? Was it slightly out of focus, or is that camera shake due to a slow shutter speed? Is the image grainy because the ISO is too high? Would the composition be improved by shooting a stitched panorama?

▲ Just owning a drone does not guarantee spectacular aerial photographs. The imagination and skills of the photographer are every bit as important. Access to some of the most beautiful spots on the planet, such as Palau, Micronesia, also helps!

COMPOSITION

From its Latin roots, the word composition literally means "putting together." Imagine yourself assembling a photograph from a collection of elements. These elements can be physical objects, such as a building in the foreground, a tree in the middle distance, and a mountain in the background; they might be slightly more abstract, such as leading lines created by objects, or patterns of light and shade; or they might involve the direction of light and use of color. In all cases these are just some of the elements that you might "put together" or use to compose a photograph.

It is not within the scope of this book to present a detailed treatise on photographic composition, but there are certain rules and tips that can vastly improve a photograph. There are also certain tricks that a drone can achieve relatively easily, which would be near-impossible for a ground-based photographer.

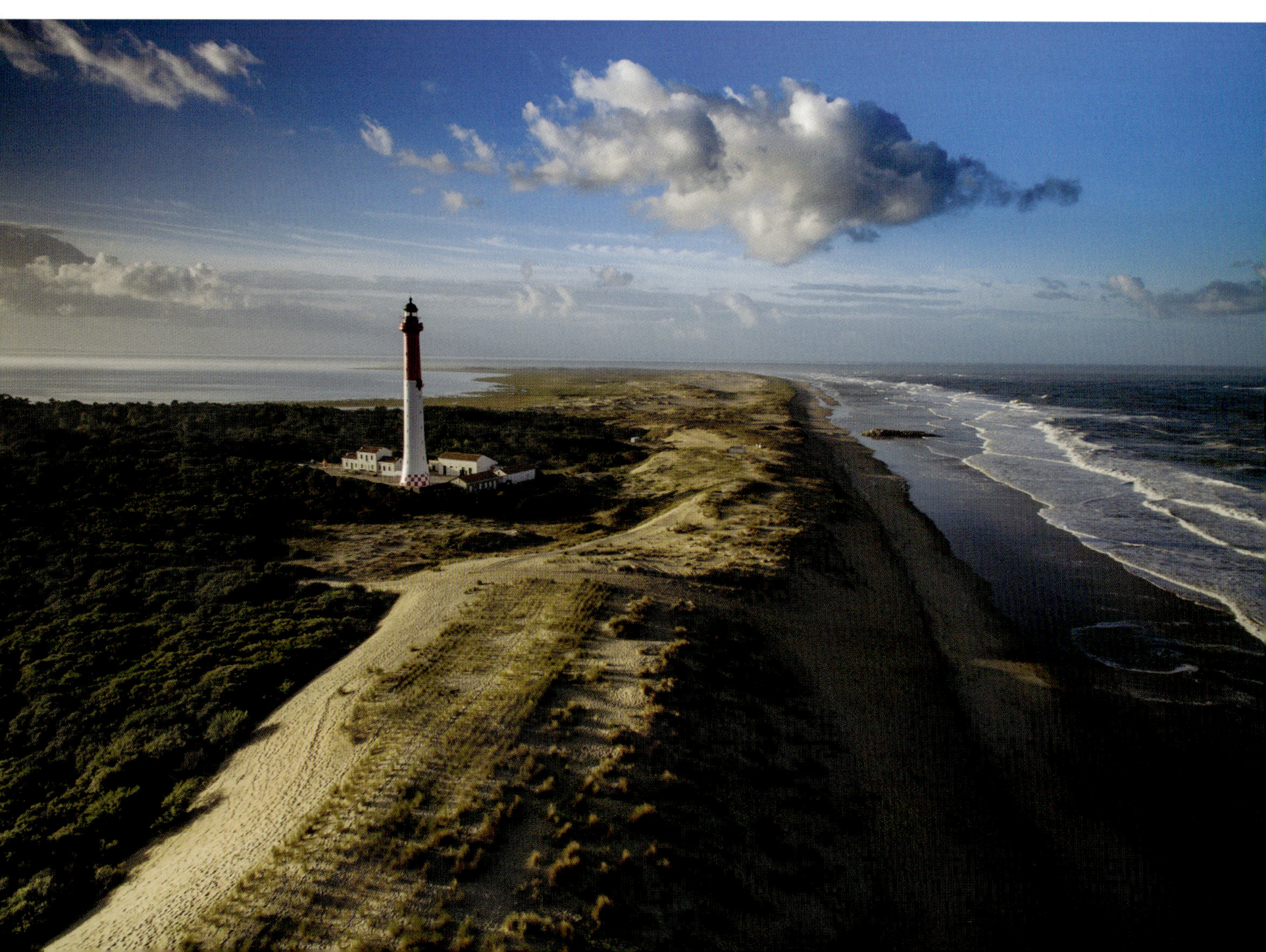

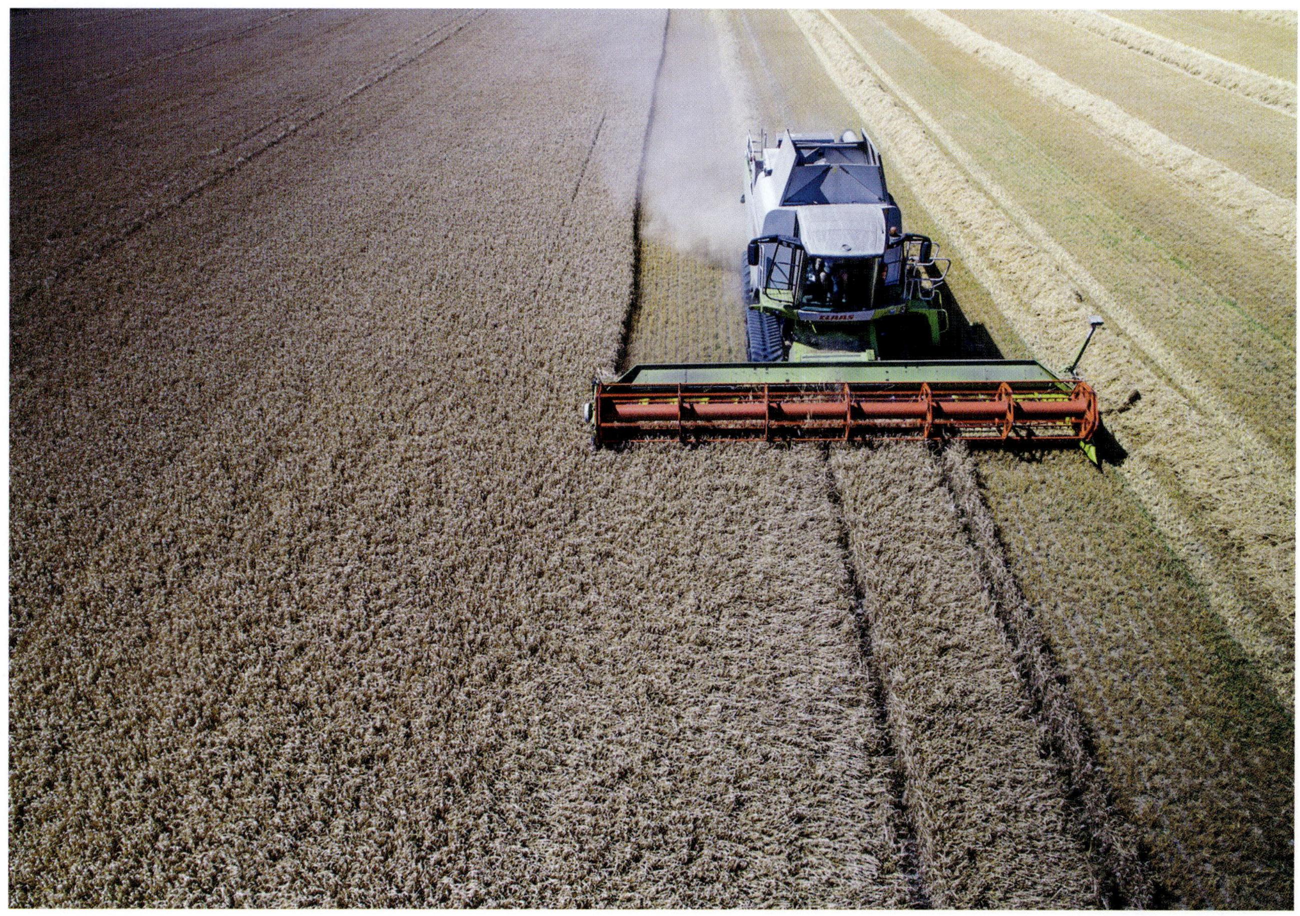

◀ The rule of thirds can help you achieve a pleasing composition, but don't adhere to it slavishly if another approach works better.

▲ Leading lines can help draw the eye into the picture, particularly with convergence due to perspective.

RULE OF THIRDS

This is a stalwart of many photography articles and books. In short, imagine two vertical and two horizontal lines dividing your frame into three equal columns and rows. The theory is that any strong vertical or horizontal elements in the scene will work best when aligned with one of these lines. Additionally, a point of interest may work best when placed on the intersection of these lines. Of course, rules are made to be broken, and you should always play around with composition and see what works for your particular shot.

LEADING LINES

Leading lines in a photograph are often said to lead the viewer's eye into the picture, so if there's any way of using these compositional elements to your advantage it often helps to make an eye-catching image. Examples of leading lines that you might find in aerial shots include the coastline, cliff tops, roads, paths, rivers, and any other line—physical or implied—that leads the eye on a specific journey.

Converging lines are particularly effective at leading the eye, especially if they take it neatly to your main subject, which is carefully placed on an intersection of the thirds lines!

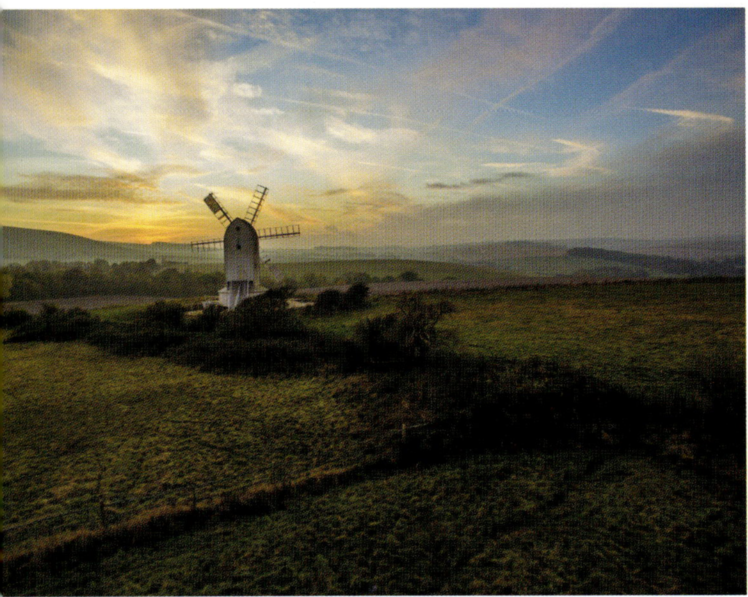

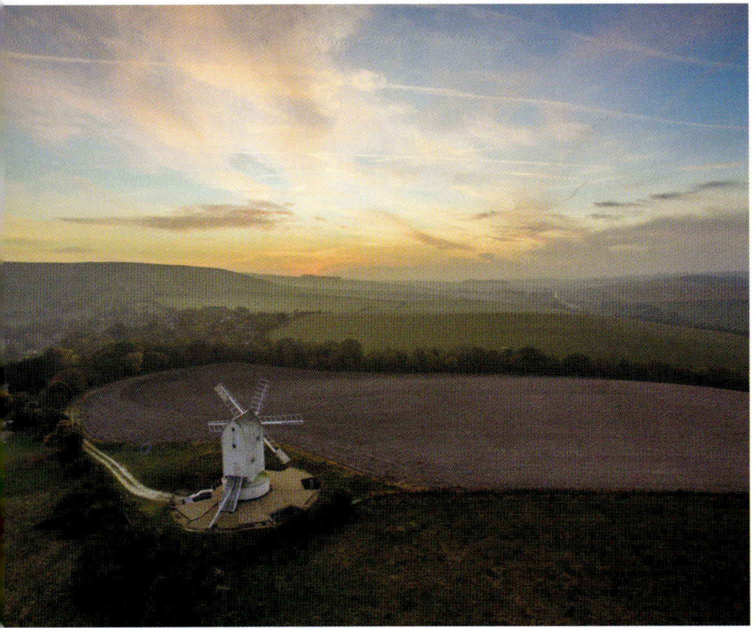

▲ A drone gives you the freedom to play around with different degrees of elevation to optimize your composition.

THE FREEDOM OF THE SKIES

From a compositional point of view, the drone photographer has one massive advantage over his ground-based brethren: the freedom to move his camera easily and quickly to almost any point in three-dimensional space (within the limits of safety and battery life of course!). When out photographing landscapes with my feet on terra firma I had often wished I could get the camera just a meter or two higher off the ground to be able to include—or exclude—a certain compositional element.

A stepladder is actually in the tool-kit of many landscape photographers for this reason, but with a drone you effectively have a fully mobile, fast-moving stepladder that's potentially up to 400ft (120m) tall, so there's much more control over how a photograph can be composed. You can get the horizon to line up with a specific element, for example; quickly shift the relative positions of foreground and background elements; or place leading lines exactly where you want them.

Finally—and this is a big one in my opinion—you have the option of a "top down" shot. Looking directly down on the world from a great height is not a perspective that is available to ground-based photographers that often, and it provides some fascinating creative possibilities. Familiar objects can appear strange, and patterns can emerge that go unnoticed from ground level, leading to stunning abstract compositions.

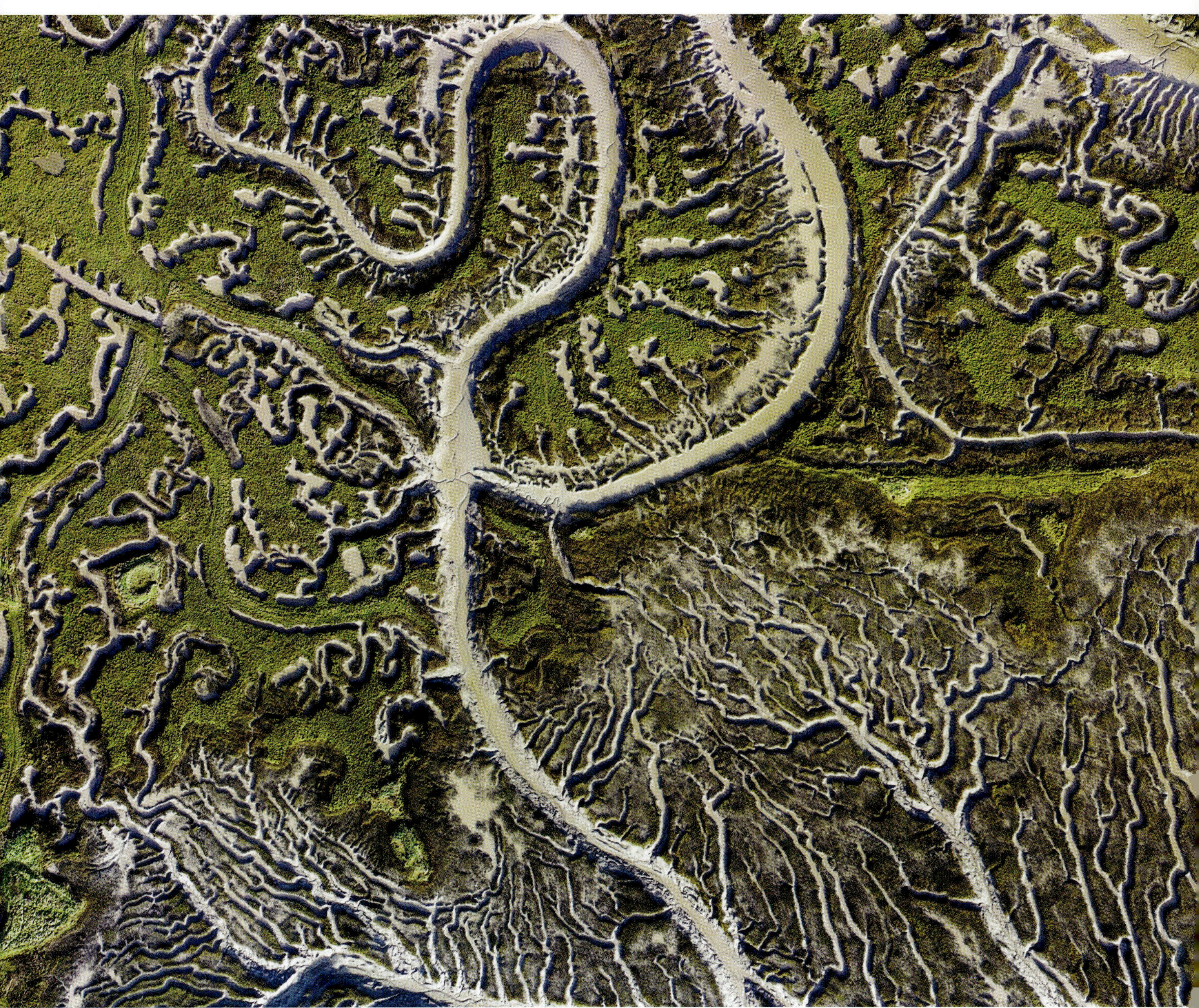

▲ Top-down shots can make for some great abstract art. It pays to spend time hunting for the best composition, and don't forget you can adjust the drone's elevation and yaw as well as its X–Y positioning.

CAMERA SETTINGS

In common with ground-based photographers, those in the early stages of mastering aerial photography might stick to the automatic settings on their camera. For the most part these will yield acceptable results, but if you want to progress you need to claw back control from the camera.

FOCUS

The degree to which you need to worry about focus will depend very much on the camera and lens you have on your drone. For the majority of users the camera will have a fixed, wide-angle focal length lens with fixed focus. That is to say it has a wide field of view and the camera's lens is locked at infinity focus. With such a setup, you do not need to concern yourself with ensuring the lens is in focus—there's nothing you can do to change it.

A slightly more advanced option is an integrated camera drone with a "tap-to-focus" system, where you tap on the screen of the phone or tablet connected to the RC transmitter to focus. Provided the subject is distant, this should only need to be done once during the flight. The deep depth of field available on a small-sensor camera will ensure that everything from a distance of around 15ft (5m) to infinity will be sharp.

Once you move to an interchangeable lens camera with a larger sensor, focusing can be a bit more critical. Depending on the setup, this is achieved using either tap-to-focus (as above), or by manually setting the focus to the hyperfocal distance before take-off (see box).

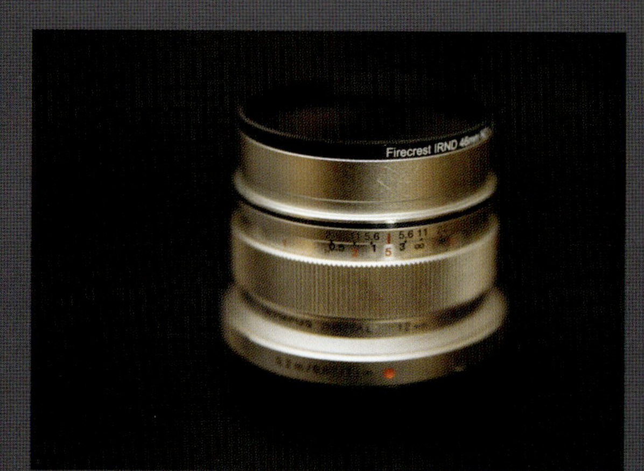

HYPERFOCAL DISTANCE

Hyperfocal distance can be defined as the focus distance that places the farthest edge of the depth of field at infinity (this is usually the sky or a distant horizon). The particular distance will vary according to your lens aperture, focal length, and sensor size.

I will not delve into the intricacies of hyperfocal distances and how to calculate them, but a quick Internet search will reveal plenty of tutorials, tools, and apps for helping with the calculations. If you are using a prime (non-zoom) lens with a clear distance scale marked on the lens barrel, you might also be able to read off the depth of field at your chosen aperture. In practice, with a bit of experimentation you will soon discover the distance at which to focus a particular lens to achieve a sufficiently deep depth of field that you don't need to worry about focus for the duration of your flight.

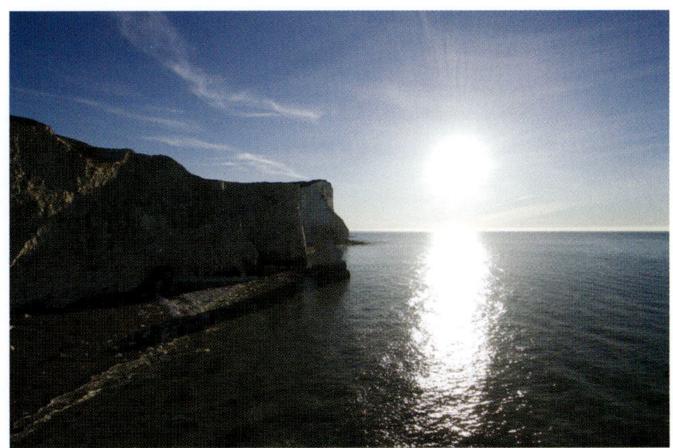
As metered: 1/1700 sec., f/2.8, ISO 200

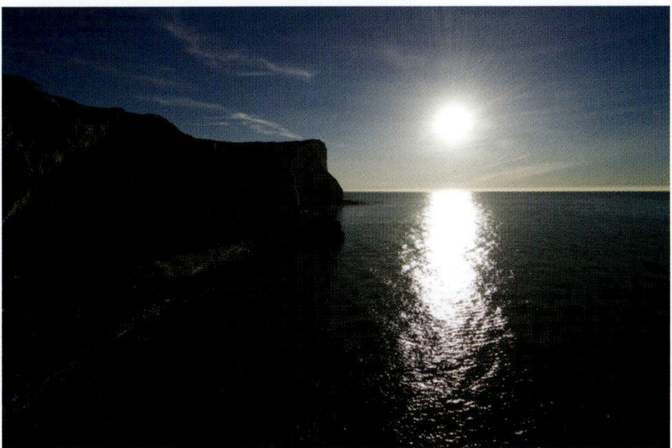
Underexposed: 1/4000 sec., f/2.8, ISO 200

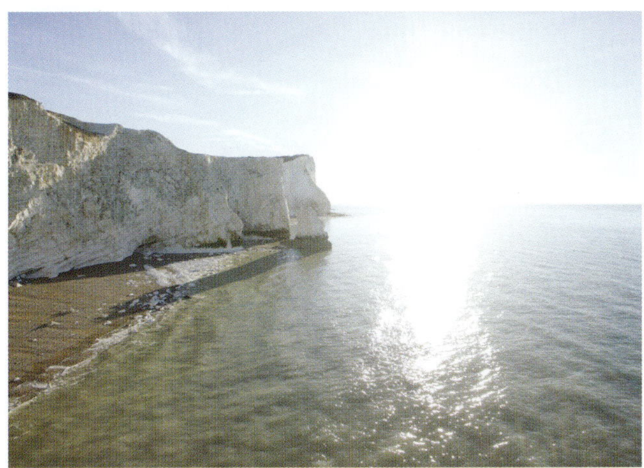
Overexposed: 1/460 sec., f/2.8, ISO 200

EXPOSURE & METERING

Having a clear understanding of the basics of photographic exposure will help you progress as a drone photographer. Exposure can be defined as the amount of light reaching the photographic plane (the camera sensor in the case of digital cameras). This is determined by both the length of time the sensor is exposed to light for (controlled by the shutter speed, which is measured in seconds or fractions of a second) and the aperture, which is the size of the variable iris inside the lens (expressed as an f-number). These two variables determine the quantity of light that reaches the sensor.

A third factor that affects how bright or dark the image appears is the ISO, which is often referred to as the "sensitivity" of the sensor. This refers to the digital amplification of the signal received by the image sensor, with higher ISO settings boosting the brightness of the image electronically in the camera.

Ultimately, the exposure settings you (or the camera) choose will be a combination of shutter speed, aperture, and ISO. This is typically a compromise between getting a shutter speed that is fast enough to avoid motion blur; an optimum aperture for picture sharpness and depth of field; and an ISO that does not introduce too much unwanted digital noise into the image. We will look at these three controls in more detail shortly.

To help you or the camera get the exposure right, most cameras have a built-in lightmeter that measures the amount of light reaching the sensor. In its automatic modes the camera will then suggest an appropriate shutter speed and aperture based on this reading (the preferred ISO is set beforehand). The lightmeter may take an average reading for the light from across the whole frame or it may take a smaller "spot" reading (or make a decision based on something between the two).

A very useful feature on many newer drones is the option to choose a point or "spot" within the scene, via the app, for the camera to take its light reading from. This means that if there is a particular part of the scene that you want to be correctly exposed, you can tap on it to set the exposure.

◀ These three shots were taken using a DJI Phantom 3 Professional. The very fast shutter speeds are a result of the camera having a fixed aperture of f/2.8 and aiming directly toward the sun.

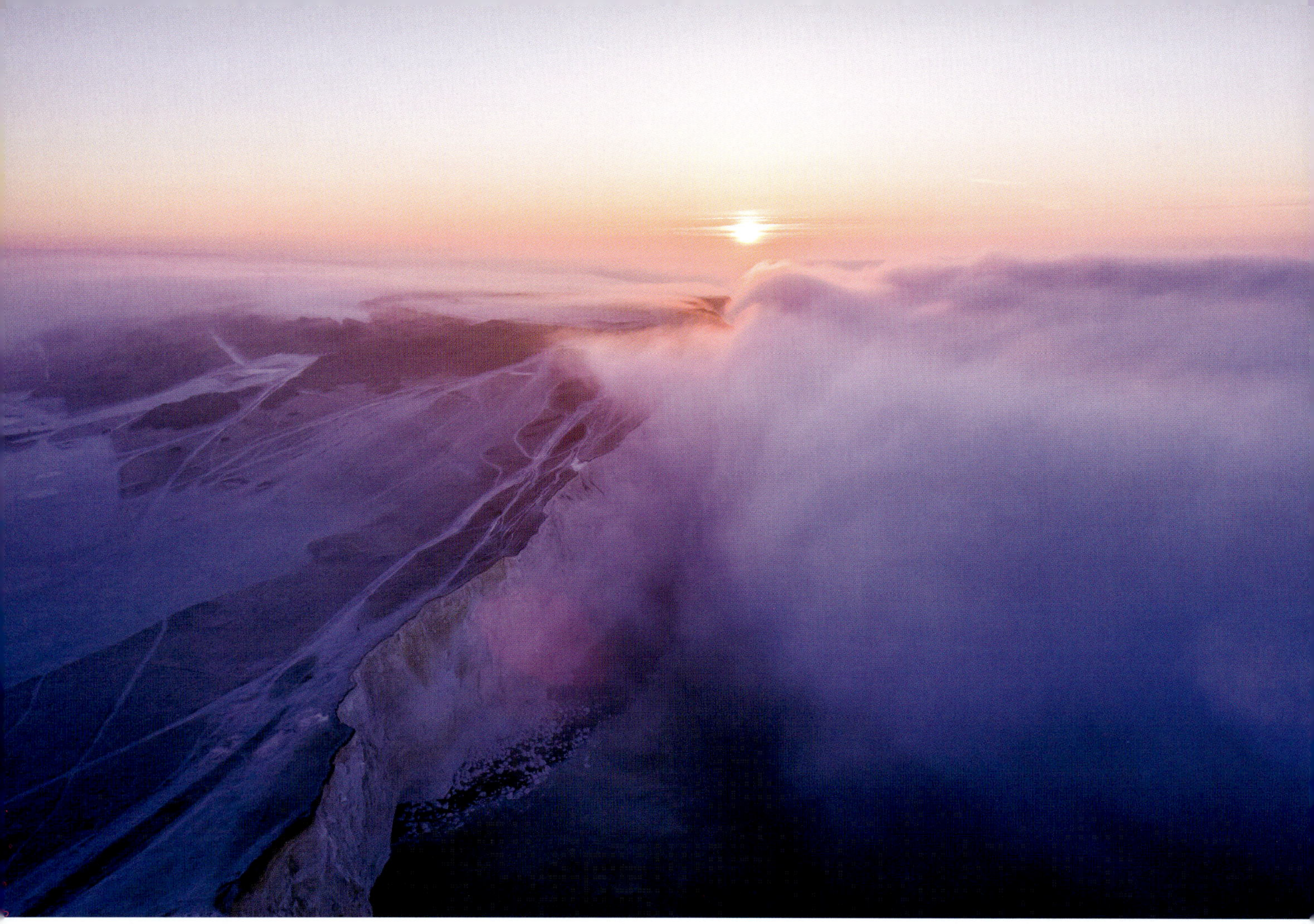

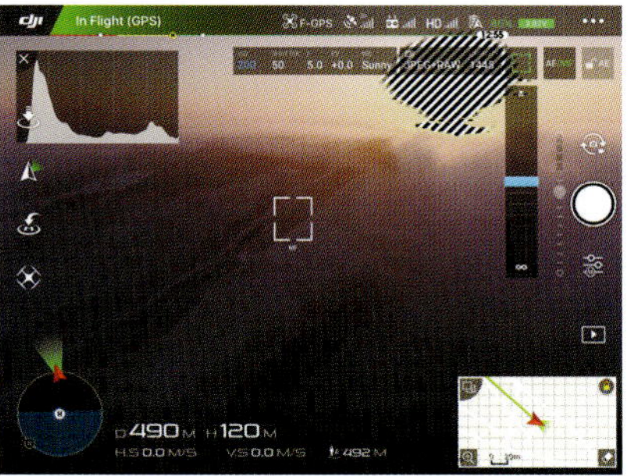

As metered: 1/50 sec., f/5.0, ISO 200

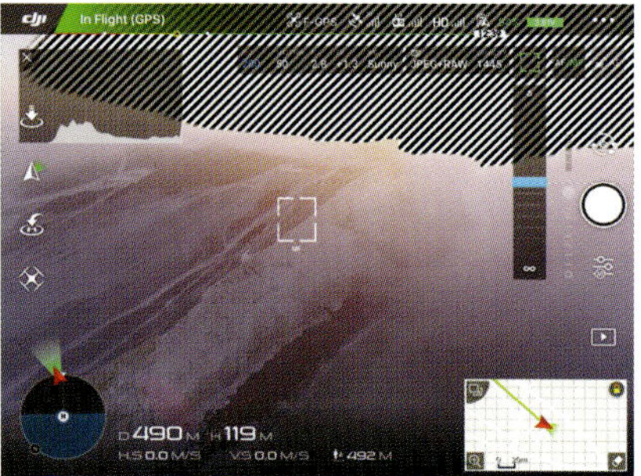

Overexposed: 1/50 sec., f/2.8, ISO 200

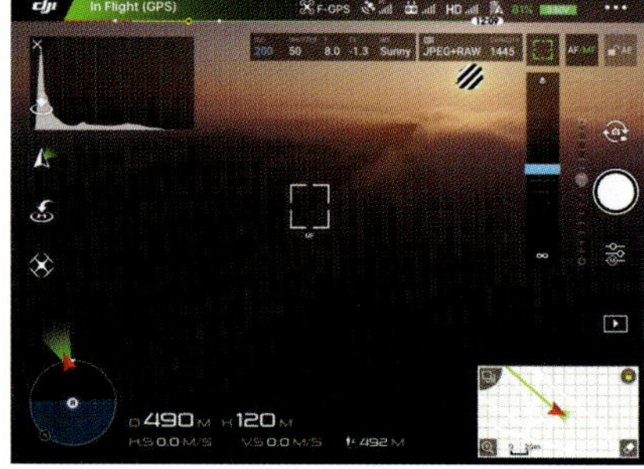

Underexposed: 1/50 sec., f/8.0, ISO 200

▲ The DJI Go App showing the same shot taken at different exposure settings (with a neutral density filter in place). Note the zebra stripes in the sky showing the overexposed areas, and the shape of the histogram for each. Overall this is a very tricky scene to expose correctly, as it contains a very bright sky and dark land. One option here would be to blend three exposures to create an HDR image (see pages 122–123), or rely on the dynamic range of the sensor and boost the shadows in post-production, while bringing back the highlights.

Histograms & Zebras

Most cameras have tools to help you check your exposure is accurate and you should be able to enable these either in the app or on the camera itself. A traditional digital camera tool is the histogram, which is essentially a graph showing the brightness values of the pixels that make up the picture: the brightest pixels are represented at the right side of the graph and the darkest pixels at the left.

Usually, you will want to keep your histogram toward the center/right of the graph and you don't want to have too many pixels off the scale at either end (this is known as "clipping"). If your histogram is colliding with the right side of the graph it means you have areas that are overexposed and detail has been lost to whiteness. This often happens with overcast skies and generally looks ugly, although small clipped areas—such as the sun if it's in frame—sometimes can't be avoided.

At the opposite end of the scale, you will have underexposed areas if the histogram collides with the left end of the histogram. Here, detail is lost to blackness and cannot be recovered.

Another tool that can help you achieve correct exposure is "zebra stripes" or "zebras." This overlays a diagonal hatching pattern over any area in the image that is overexposed or close to being overexposed (the exact cut-off point can often be changed in the tool's settings).

Whichever tool you use, if you are not happy with the exposure, your camera should allow you to apply exposure compensation to brighten or darken the shot as necessary, before you reshoot.

◀ This is a tricky exposure situation as the shot contains the rising sun and areas of land that are still in deep shadow. This is where exposure aids such as histograms and zebras come into their own, helping you avoid blown highlights or lost shadow detail.

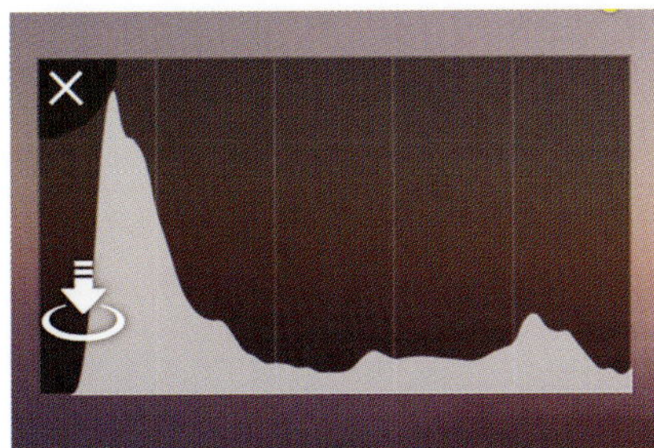

▲ The camera's suggested exposure results in an overexposed area around the sun, as indicated by the zebra stripes. The histogram shows a big peak at the left side, which corresponds to the dark pixels in the bottom ⅔ of the image (the land). The smaller peak on the right is the brighter sky.

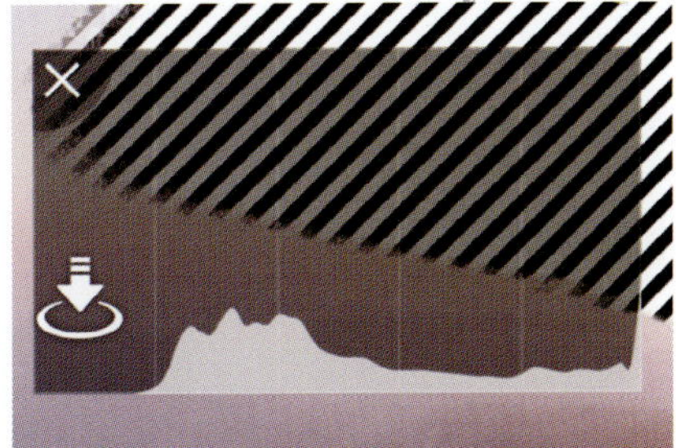

▲ Increasing the exposure by 1⅓-stops to brighten the land means that most of the sky is overexposed (as shown by the zebra stripes). In the histogram, the land pixels have moved closer to the center of the histogram, but the sky pixels have moved off the right side of the graph.

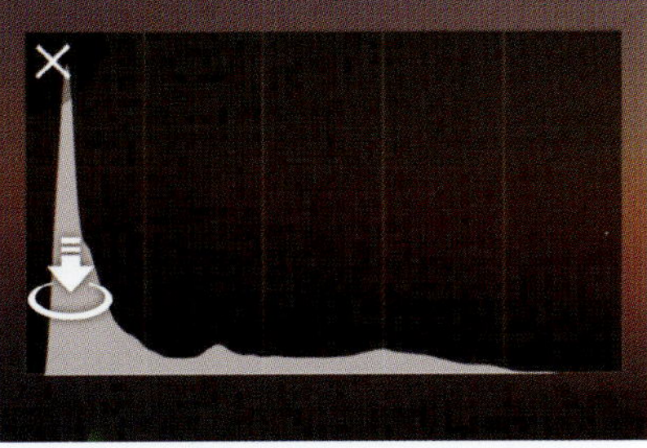

▲ Reducing the exposure by 1⅓ stops means that only the ball of the sun is overexposed, but the land pixels are now very close to the left edge of the histogram, indicating they are very dark.

SHUTTER SPEED

The longer the shutter is open for, the more light reaches the sensor, but as any keen ground-based low-light photographer will know, longer shutter speeds run the risk of camera-shake—if the camera moves while the shutter is open, the picture can appear blurred.

You might think this would be a serious problem on a small drone with vibrating motors that is often being buffeted by the wind, but in practice modern gimbals with rubber dampers to absorb vibration do a remarkable job of keeping the camera steady. I've seen pin-sharp drone photographs take at night with exposures in the region of 5 seconds (although this may take a certain degree of luck, persistence, and calm weather!).

For still photographs taken in broad daylight, the shutter speed will be fairly fast, and under these conditions camera shake is rarely a problem. As a rule of thumb if you are shooting from a small drone with a wide-angle lens and a good gimbal you get sharp photographs at a shutter speed faster than 1/50 sec. Depending on the conditions you may be able to go slower than this.

MOTION BLUR

Sometimes the blur created by using a slow shutter speed can be desirable. Motion blur can give the illusion of speed to a still photograph. Imagine an image of a cyclist, where the subject is sharp, but the background is blurred horizontally—it immediately gives you a sense that the subject is in motion.

This type of image can be achieved relatively easily with a drone. You track your subject in order to keep them centered in the frame (this requires a certain amount of skill) and photograph them with a relatively slow shutter speed (of 1/30 sec. or slower). In order to achieve this slow shutter speed you may need to shoot in low light (at dusk or pre-dawn) or use a strong neutral density (ND) filter on the lens to reduce the amount of light reaching the sensor.

▼ **Deliberately including motion blur will create an impression of speed. For this shot, an ND filter was used to extend the exposure time while the drone was tracking the cyclist.**

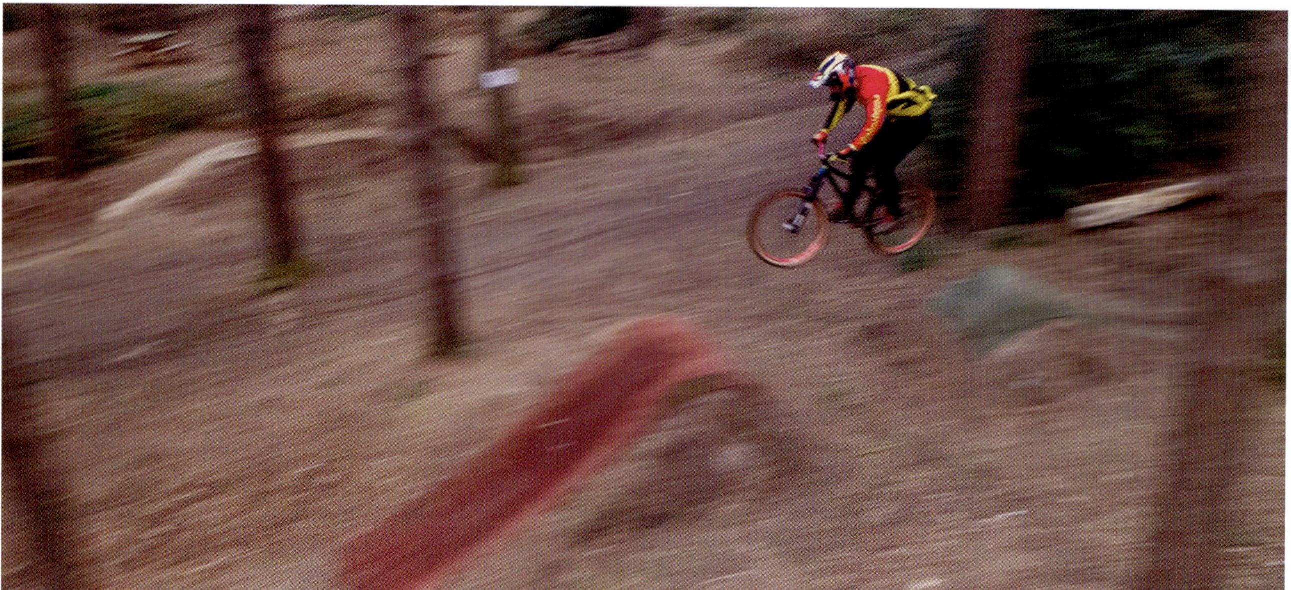

1/50 sec., f/5.6, ISO 200 (with ND32 filter)

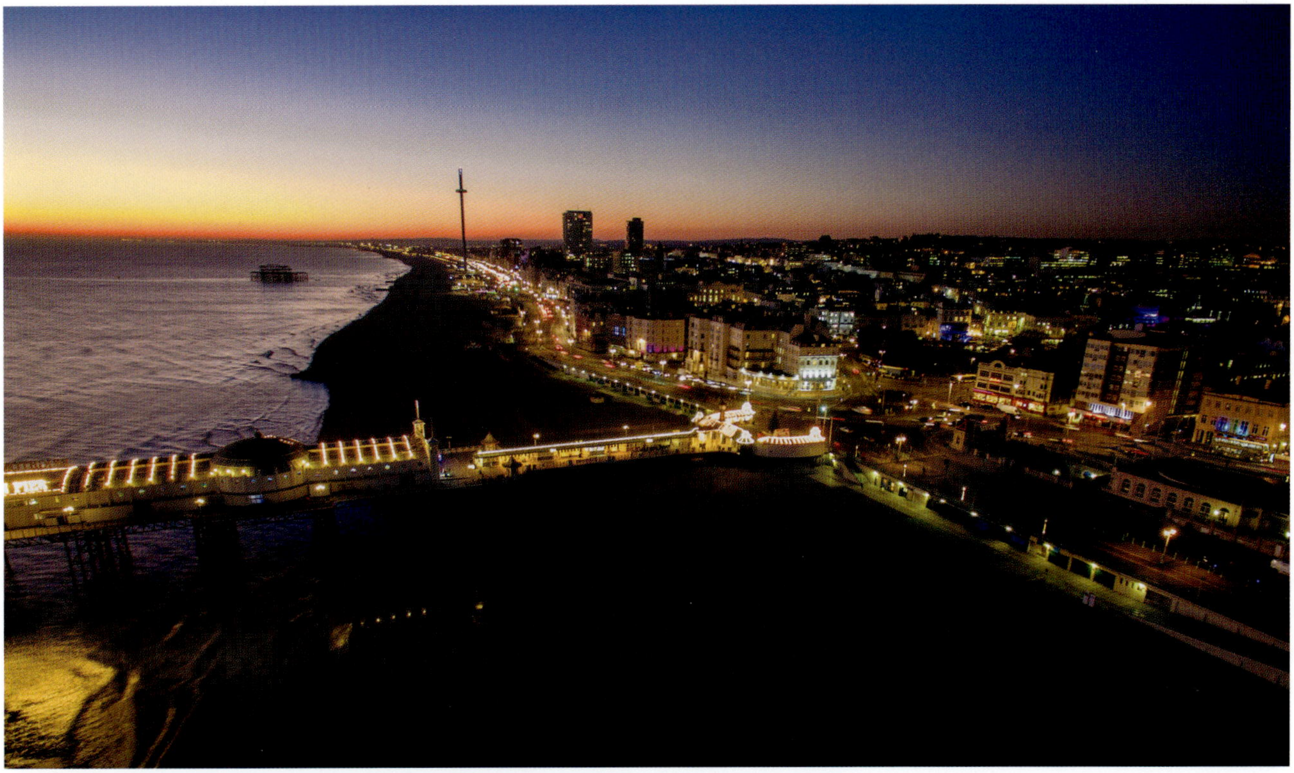

No motion blur, as aircraft was steady: 1/2 sec., f/2.8, ISO 200

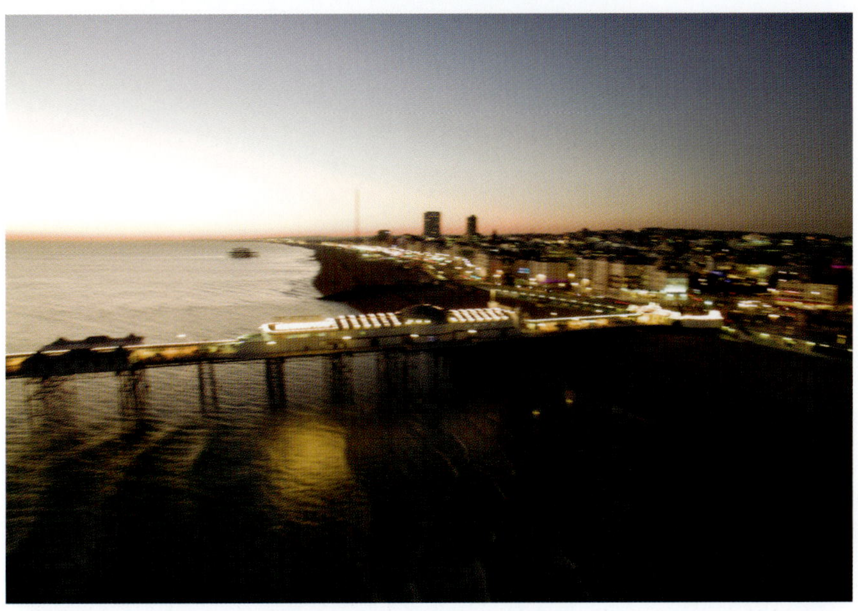

Motion blur caused by drone movement: 1/2 sec., f/2.8, ISO 200

◀▲ When the light level gets low you will need to use a slower shutter speed. However, you may also need several attempts to get a clear shot, particularly if it's windy, as the drone needs to be steady. You should also ensure you have removed any ND or polarizing filters that would extend the exposure time.

APERTURE

All lenses have an aperture, which is a hole through which light passes on its way to the sensor. The majority of small drones with integrated cameras have a fixed aperture (often f/2.8 or f/2.2), so you don't need to worry about setting it: image brightness is changed using shutter speed and ISO.

However, larger drone-based cameras (especially those that use interchangeable lenses) will have a variable aperture that can be set either in-flight via an app or manually on the camera, prior to take off. Most lenses of this type have an optimum aperture at which they produce the sharpest images, which is usually in the middle of the range (typically around f/5.6–f/11). Beyond this, softness or diffraction can decrease image quality, although this is often not significant enough for most people to worry about.

The other effect of the lens aperture is that depth of field will *decrease* with a wider aperture setting (smaller f/number). In land-based photography this can be a very creative tool, enabling you to make your subject stand out from its background. This is less relevant with most drone photography, though, as depth of field *increases* with camera-to-subject distance. As the main subject will typically be at a distance of more than 30ft (or 10m) when you're using a drone, the depth of field is usually such that everything appears sharp, even when a fairly wide aperture setting is used.

My usual rule of thumb is that if I have plenty of light I will shoot in the optimal range of f/5.6–f/11. If the light level drops I'll increase the ISO, but only to a maximum of ISO 800—at that point I will start to open up the aperture, all the time trying to maintain a reasonable shutter speed. However, the strategy that works best for you will depend on your camera, so it is worth experimenting to find the optimum settings.

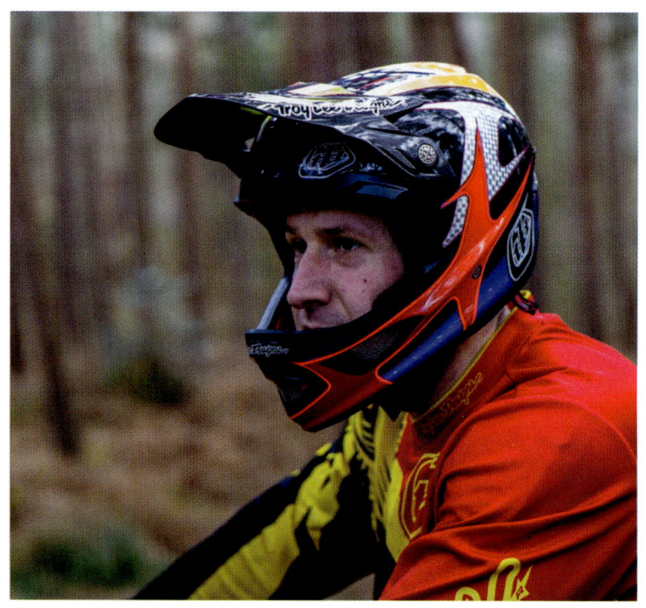

f/1.8

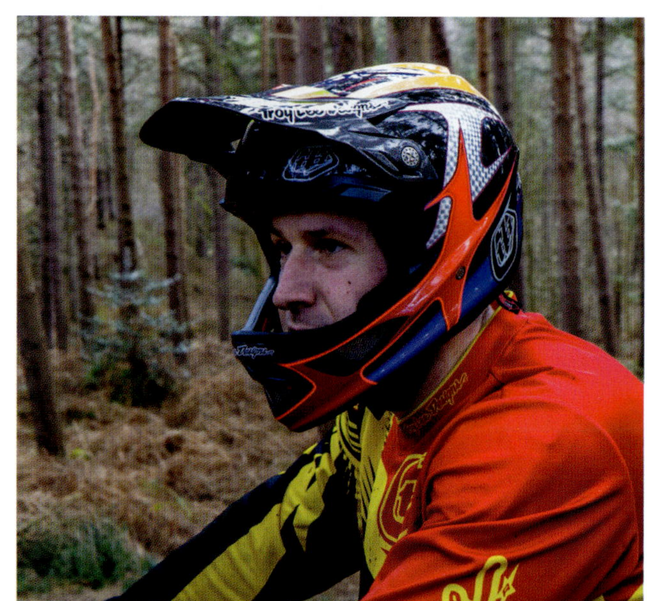

f/11

▲ These shots were taken using a 90mm equivalent focal length lens. Changing the aperture has had a significant effect on the depth of field: note the sharpness of the background in both images.

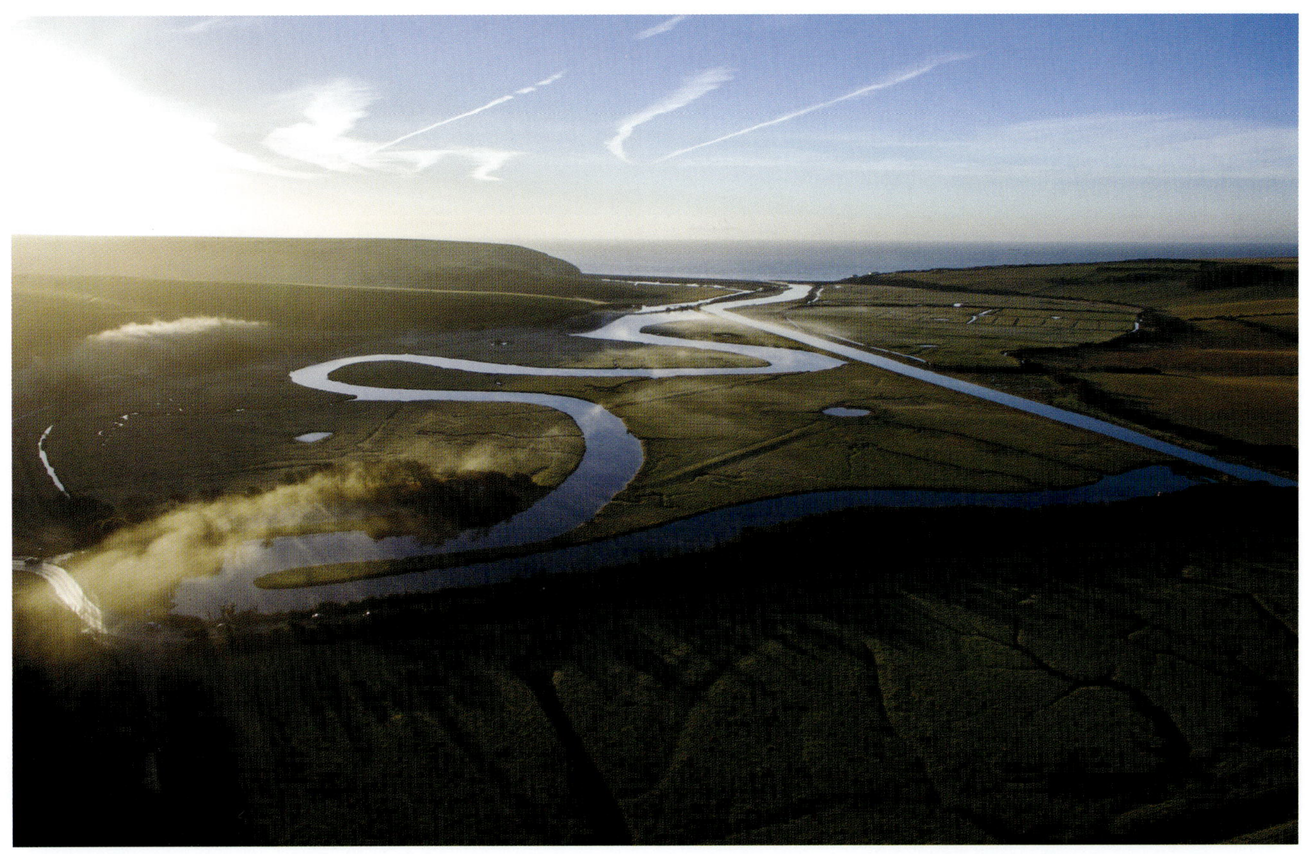

▲ With a wide-angle lens and relatively distant subject matter, the depth of field will be deep at all aperture settings.

DEPTH OF FIELD

With any camera there will be a point (or more accurately a distance) at which the image is at its maximum sharpness—this is the distance where something will be "in focus." However, there is also a certain distance in front of and behind that point that also appears acceptably sharp: this distance range is known as "depth of field."

When photographers refer to a "shallow" depth of field, they mean there is a very narrow range that is "in focus." For example, in a portrait the subject might be sharp, while the background is out of focus, or in a more extreme example, just the subject's eye will be in focus, while their nose and the ears, and everything beyond is blurred.

The depth of field in a photograph depends on a number of factors: the aperture the lens is set at, the focal length of the lens, the camera-to-subject distance, and the sensor size. Wider apertures (smaller f/numbers), longer focal length lenses, closer camera-to-subject distances, and larger sensors all lead to a shallower depth of field.

However, the control you have over depth of field as a drone photographer depends to a large degree on your equipment. The cameras on most drones—particularly the small ones—have fixed apertures, relatively wide-angle lenses, and a small sensor. Combine this with a (usually) large camera-to-subject distance and the depth of field will be deep, with everything appearing in focus. It is not until you start using a camera with a larger sensor, longer focal length lens, and variable apertures that you can start making creative use of shallow depth of field effects.

ISO

The ISO setting indicates the amount of digital gain that is applied to the signal created by the light hitting the sensor and provides an additional way for the photographer to adjust the brightness of the image.

Every camera has a "base" or "native" ISO, which is often ISO 100 or ISO 200. Any setting above this has a digital signal boost applied to it, which will entail a more or less noticeable loss in image quality in the form of digital noise or grain.

The extent to which this noise becomes a problem is highly dependent on the camera. On small drones with integrated cameras, high ISO noise may start becoming noticeable at ISO 800 or above, whereas modern DSLRs can often shoot very clean images at ISO 6400 and beyond. It's a good idea to experiment with your particular camera to see what ISO levels it can handle.

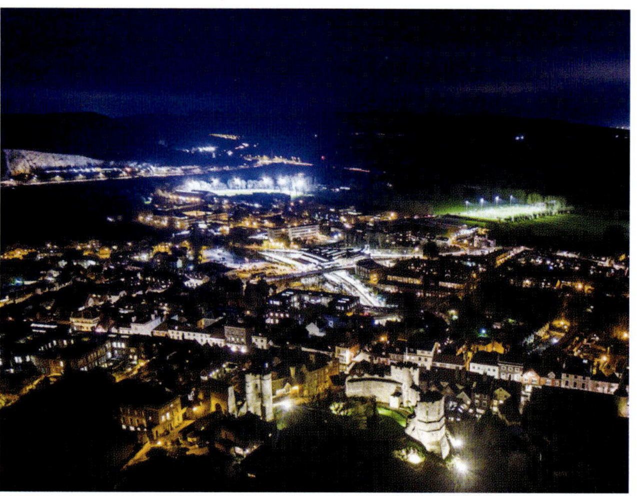

▲ If you're shooting in low-light conditions your choice of ISO needs to be balanced with the shutter speed: if the shutter speed is too low, you risk introducing camera shake .

▼ Taken with the ISO set at 100, the dark areas are still relatively clean in this detail. However, you need to make sure this doesn't mean you're using a shutter speed that will introduce camera shake**.**

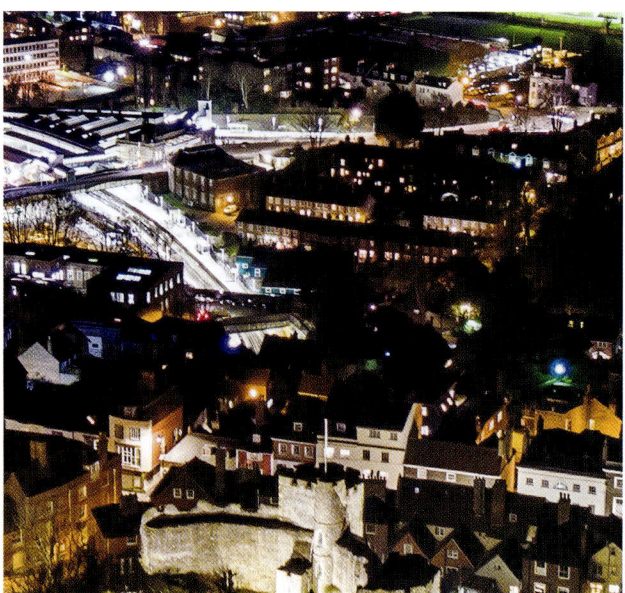

▼ Here, the ISO was increased to ISO 1600, which gave me a much shorter exposure time. While this reduced the risk of camera shake it has led to unacceptably high noise levels.

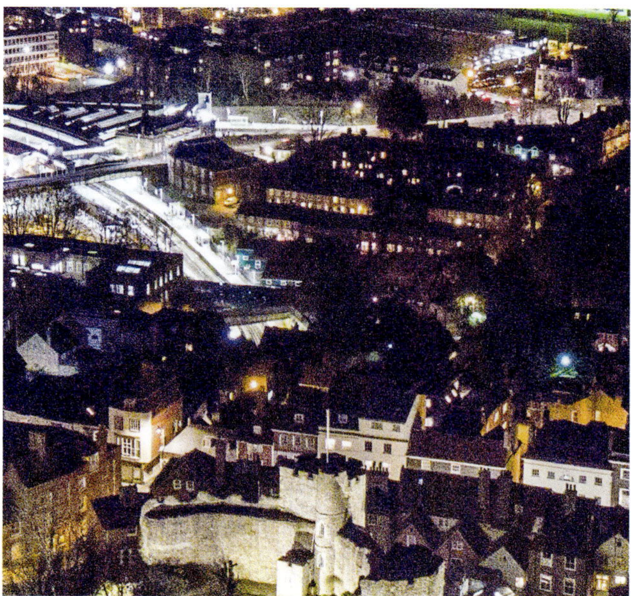

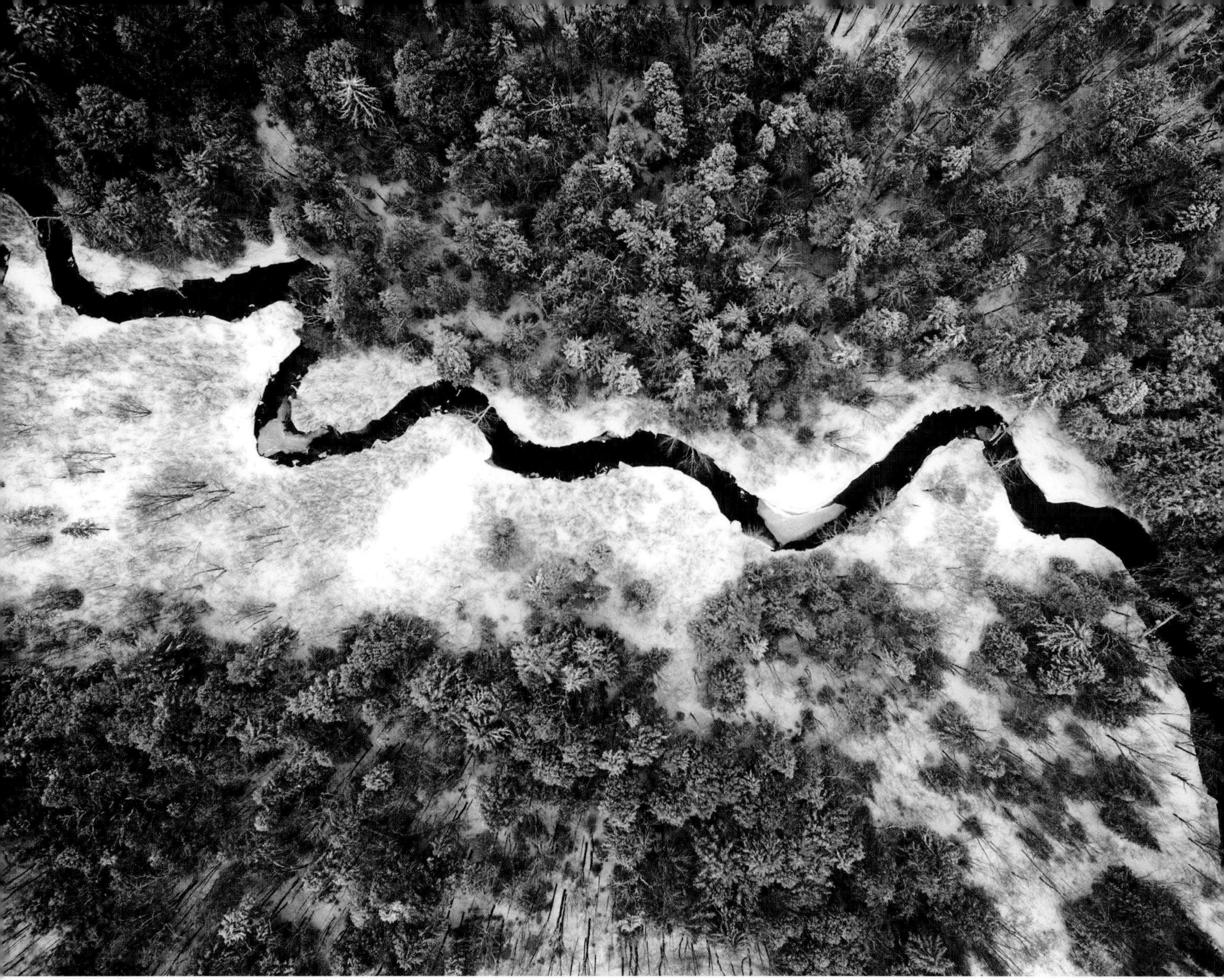

▲ In good light there is a very simple rule: use the lowest ISO you can. As long as the shutter speed isn't too slow, there's no risk of camera shake, so your low ISO setting will ensure you minimize noise and get the best image quality, as in this shot by Andy Deitsch (profiled on the following pages).

PROFILE: ANDY DEITSCH

Andy Deitsch is a US-based aerial, underwater, and wildlife photographer. He has always been interested in photography, but it wasn't until 2011—when he had the opportunity to borrow an underwater camera—that he started to take it more seriously.

After a number of underwater photography workshops, what started as a hobby soon became a career and Andy increasingly found himself on assignment in remote regions of the world. He realized that drone photography would be a great way to capture the natural beauty of these locations during surface intervals and now takes a small drone with him on all his travels, in pursuit of novel camera angles.

▼ This shot was taken in Palau, in the Western Pacific Ocean, while Andy was on a diving expedition. Because of limited space on their small boat, Andy found it best to hand launch and hand catch the drone.

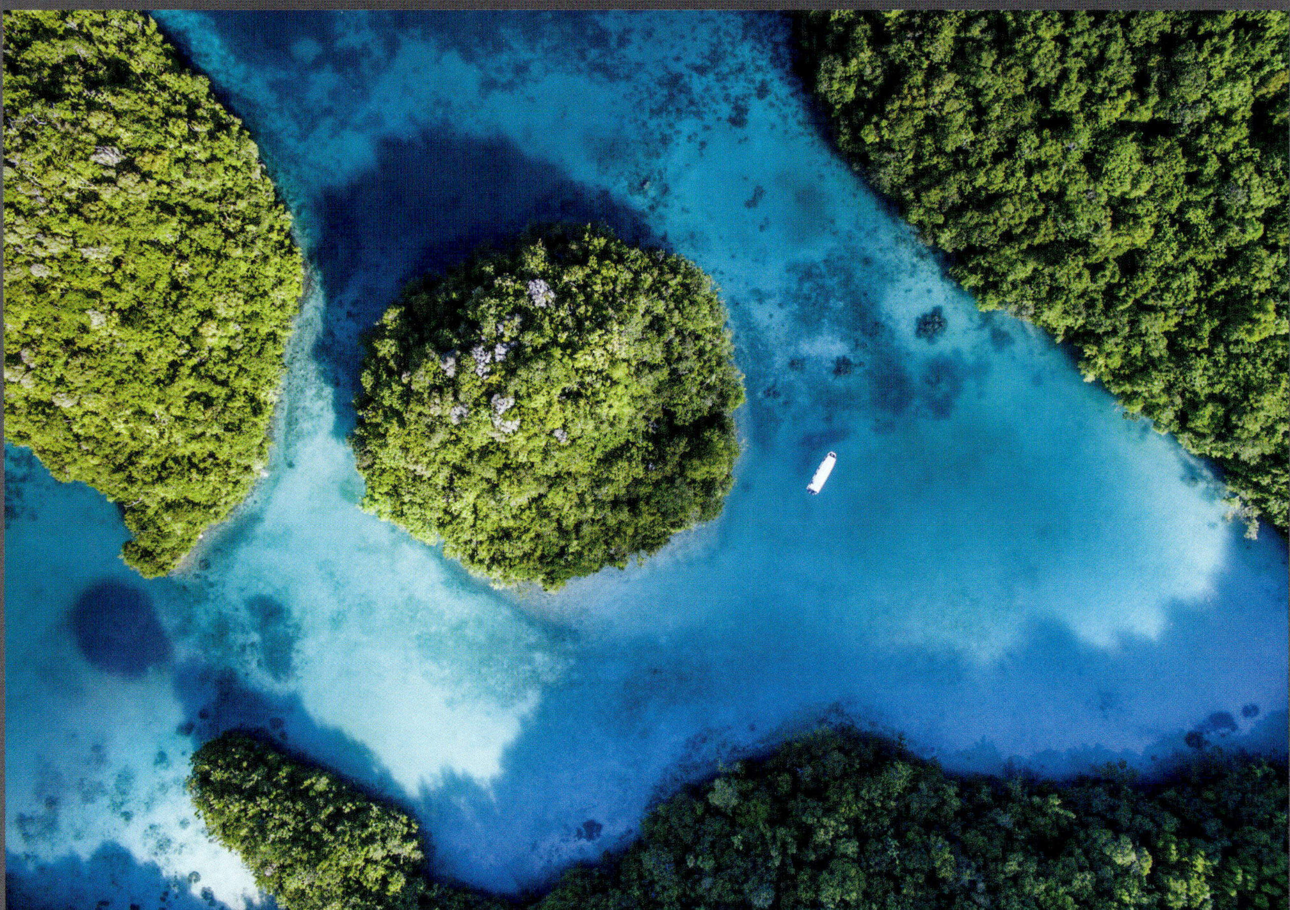

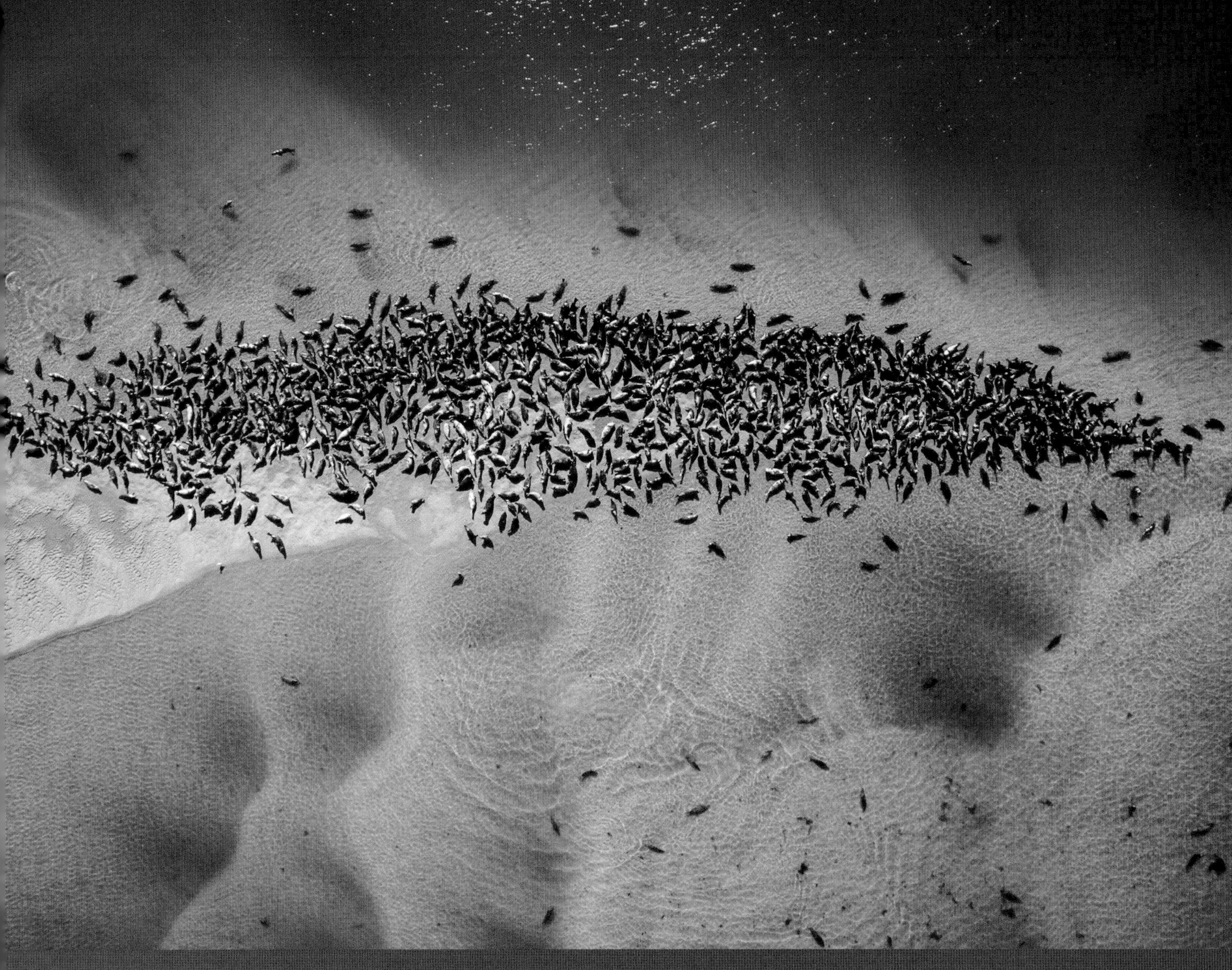

▲ This group of seals was sitting on a sandbank about 2000ft (600m) from shore at Chatham, USA, and using a drone was the only way to reach them. Andy took this image with a DJI Phantom 3 Professional.

To maximize the quality of his images, Andy will often take several overlapping shots and stitch them together, rather than flying a distance away from the subject to get it all in the frame. However, unlike a stitched panorama (see page 126) he will shoot both horizontally and vertically, using yaw to pivot the drone in place for the horizontal shots and the gimbal to move to the next row. This results in a final image that has more pixels and therefore more detail.

LENS CHOICE

Small drones with integrated cameras usually have a fixed, non-changeable lens with a wide or very wide angle of view (typically in the region of a 70°–90° horizontal viewing angle).

For the majority of users this is a great way of getting an epic view from the air, as one of the effects of a "wide-angle" lens is that subjects look farther away than they actually are. This means you get a greater sense of height (without actually being too high) and can fit a lot of the landscape below into a shot.

The downside is that if you have a particular subject to photograph—a person or a car, for example—the drone has to be relatively close for them to be recognizable. This, of course, has safety implications.

GETTING TIGHTER

Drone users who have interchangeable lenses on their camera have the choice to use a "less wide" (telephoto) lens. This could be a prime lens or even a zoom, but gives you the option of tightening the view on a particular subject while maintaining a safe working distance.

Although you can shoot the same subject from very close with a wide-angle lens or from a greater distance with a telephoto lens, the compositional effect will be very different with each approach. A wide-angle lens used close gives a more distorted impression of the subject and will include a much wider background, while the tighter lens will include a smaller angle of background and background objects will appear closer to the subject. It is also worth noting that the effects of camera shake will be exaggerated on a longer focal length lens, so you will need to use faster shutter speeds (typically 1/250 sec. or faster).

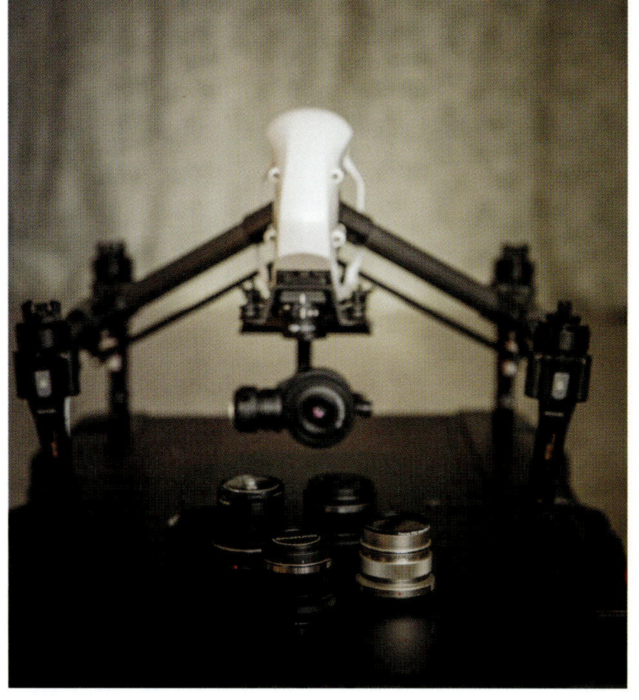

◀ A DJI Inspire Raw along with a range of interchangeable lenses. It's by no means essential, but having the ability to shoot at different focal lengths gives you more options with composition and the look of your images.

▶ These images were taken using different focal length lenses, keeping the castle tower in approximately the same position in the frame, at roughly the same size. Note the relative appearance of the background features in the two images: the background appears much "closer" to the castle in the photograph taken with a 90mm equivalent telephoto focal length.

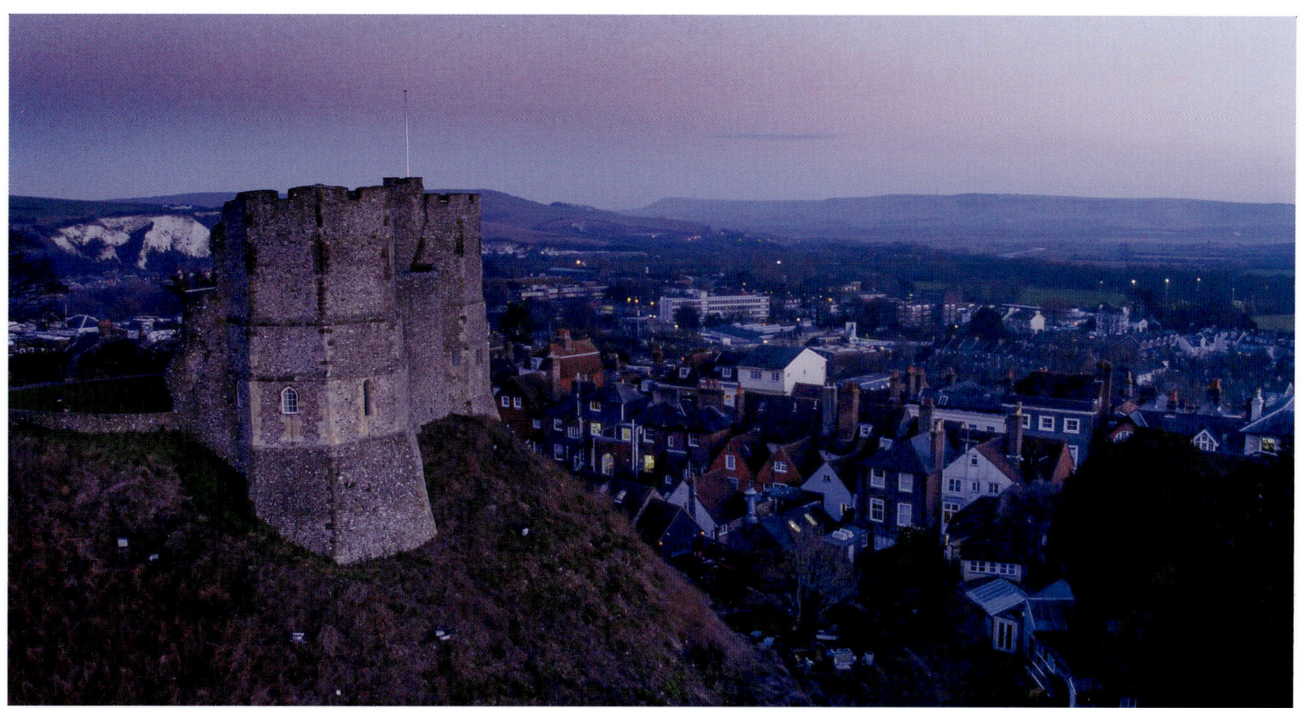

24mm equivalent wide-angle focal length

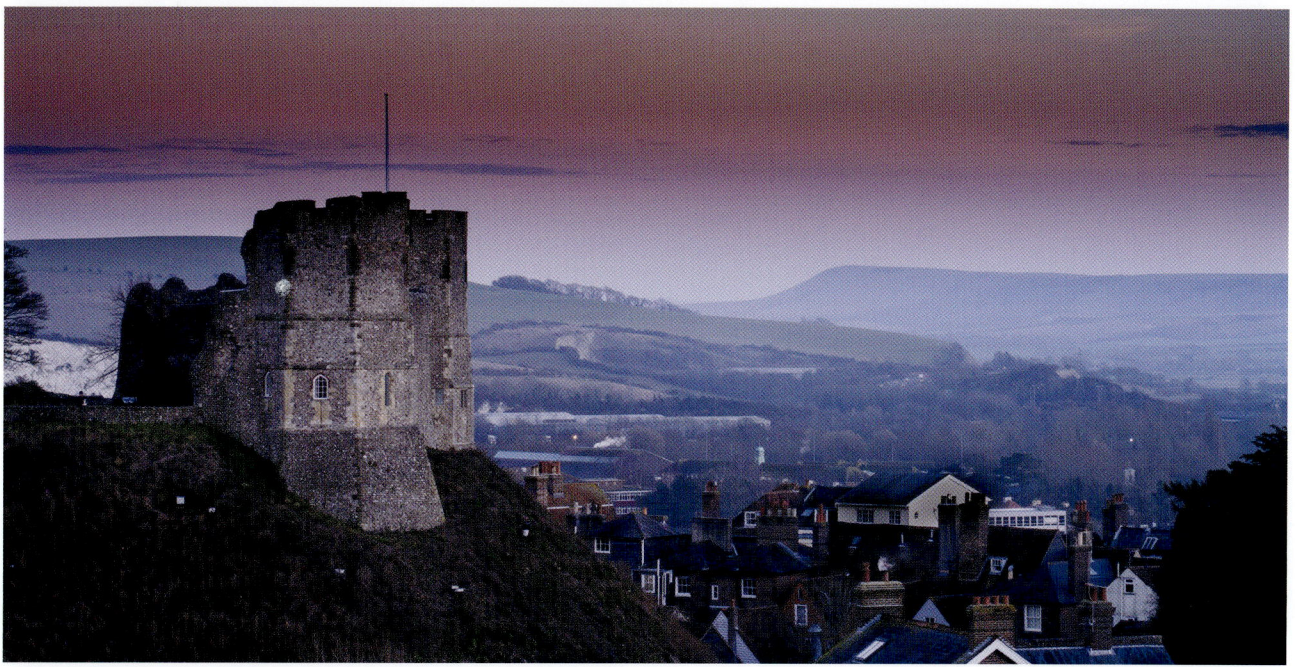

90mm equivalent telephoto focal length

IMAGE FORMATS

Many drone cameras now give you the option to shoot in either a compressed JPEG file format or a Raw format, such as DNG. The JPEG format was developed primarily to save space on memory cards and if you want to take photographs and share them immediately with minimum fuss, this is the format to stick to. JPEG images have a "baked in" look, whereby you can select a picture profile before you take a shot, which will include values for parameters such as contrast, color saturation, and sharpening. Although these can be changed using image-editing software, the quality of the image can degrade as a result.

Raw photographs give you far more latitude to adjust them later on your computer or tablet, and this is the recommended format if you want to make the most of your aerial photography. If you shoot a Raw image you will have preserved the maximum amount of information possible from the camera sensor, which gives you the greatest flexibility when you process the image. Unlike a JPEG, you can make more and larger adjustments to parameters such as exposure, white balance, and color saturation without degrading the image.

In either case, it makes most sense to choose the highest possible resolution (most pixels) if you have the option. You should also choose the camera's native aspect ratio (often 3:2 or 4:3) rather than an in-camera crop (such as 16:9), which will use fewer pixels and therefore create lower resolution images. Should you later decide to crop the image this can be done on your computer or tablet.

▶ Every camera has a "native" aspect ratio, which has the same proportions as its sensor. However, you can usually set alternative aspect ratios, which alter the frame shape by cropping the image.

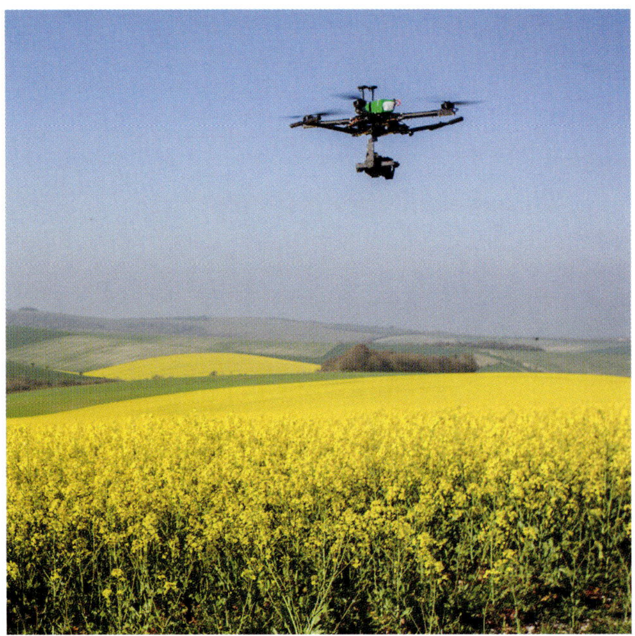

1:1

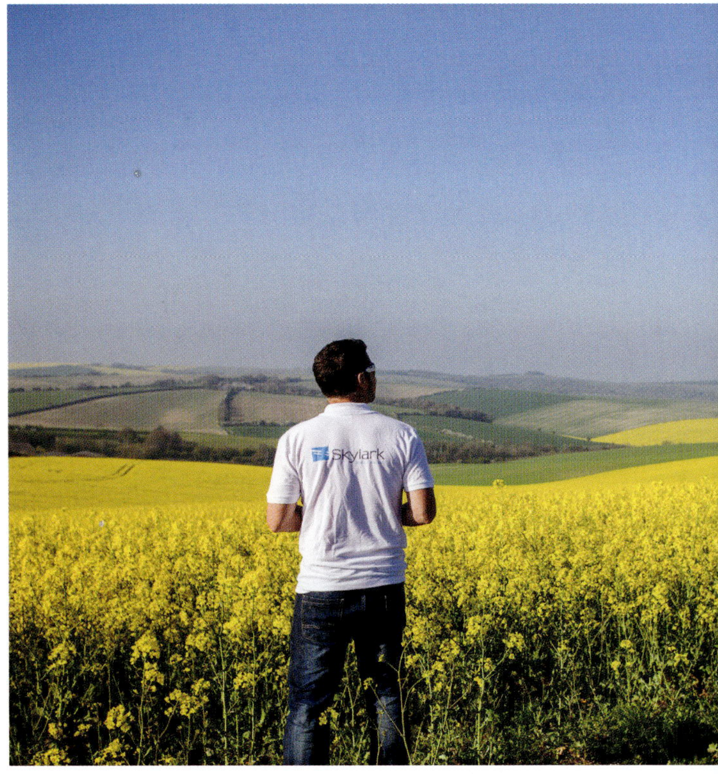

3:2

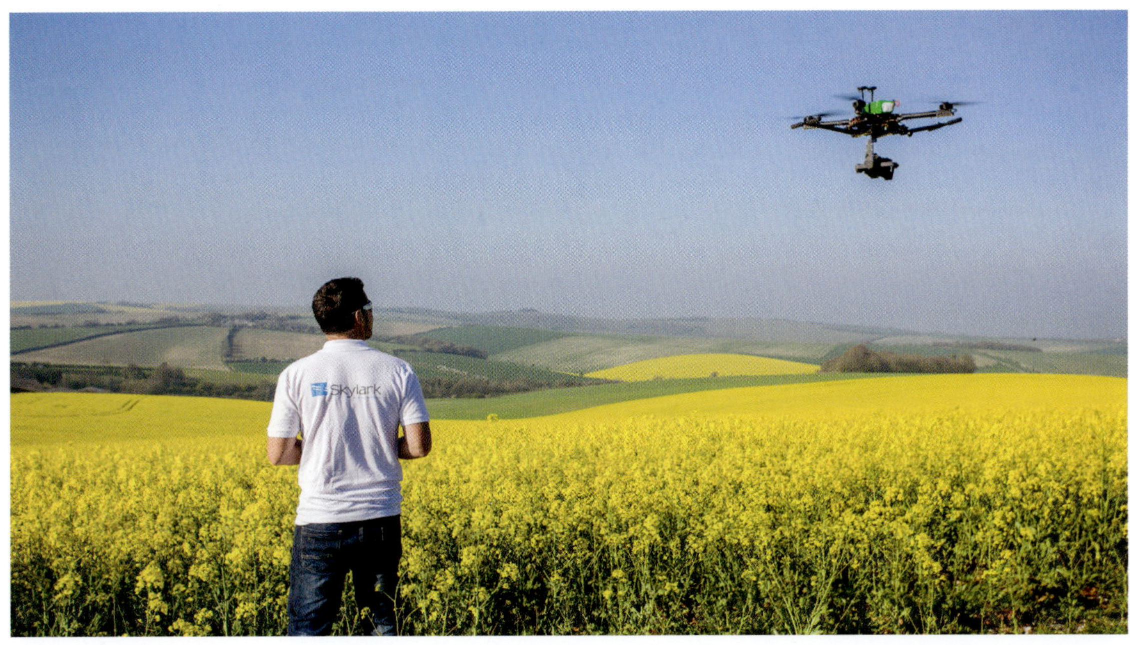

16:9

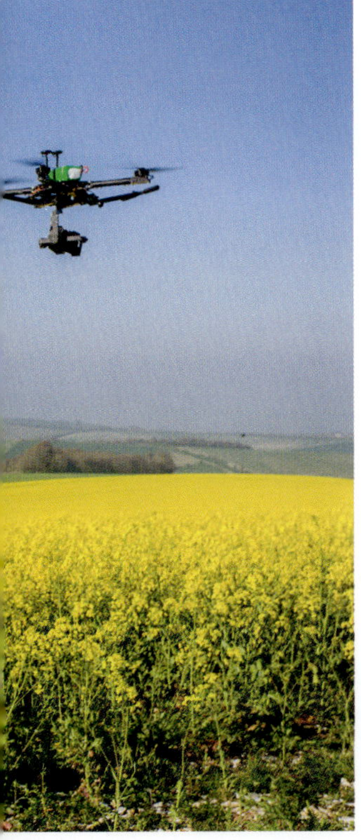

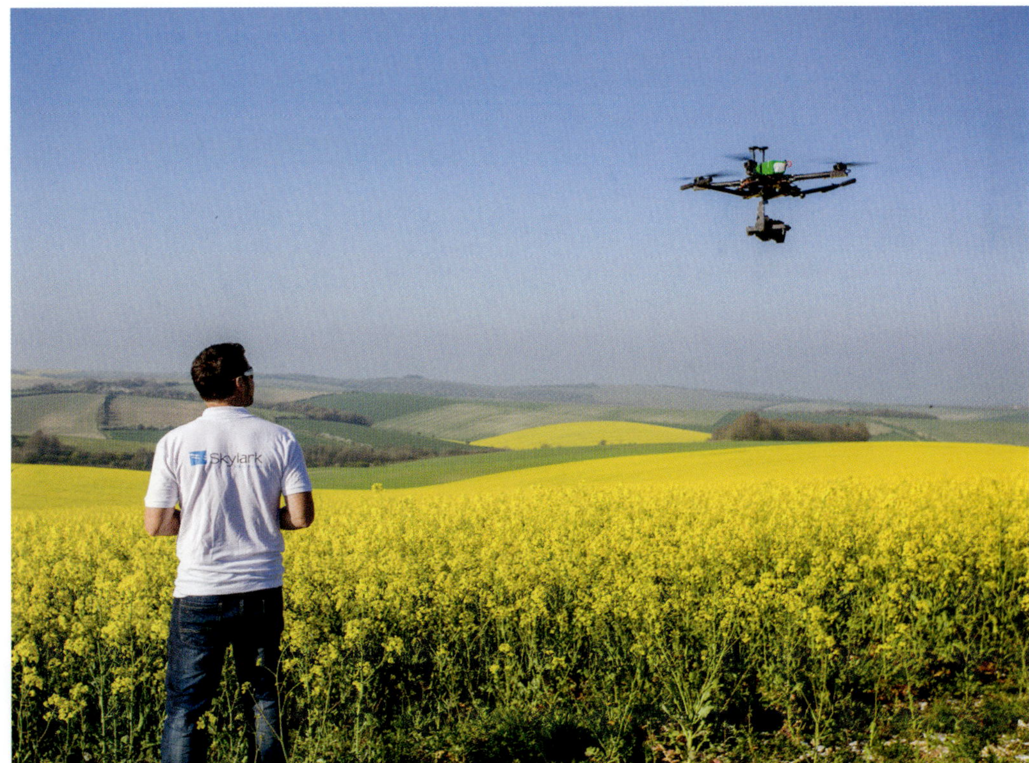

4:3

POST-PROCESSING

Although free and low-cost image-editing programs are available, it is usually worth spending a little bit more on good software that will handle Raw picture formats. Adobe Lightroom and Photoshop are very popular, but there is also a number of cheaper alternatives.

This book will not go into detail on post-processing, as there are plenty of good resources, and online tutorials on the subject out there. Most software will give you access to a number of key controls: the ones I use most are Exposure, White Balance, Highlights & Shadows, Clarity, and Sharpening. Local adjustments such as brushes, cloning and healing tools, and grad filters are also useful for removing or toning down unwanted elements, and making other localized adjustments to the image.

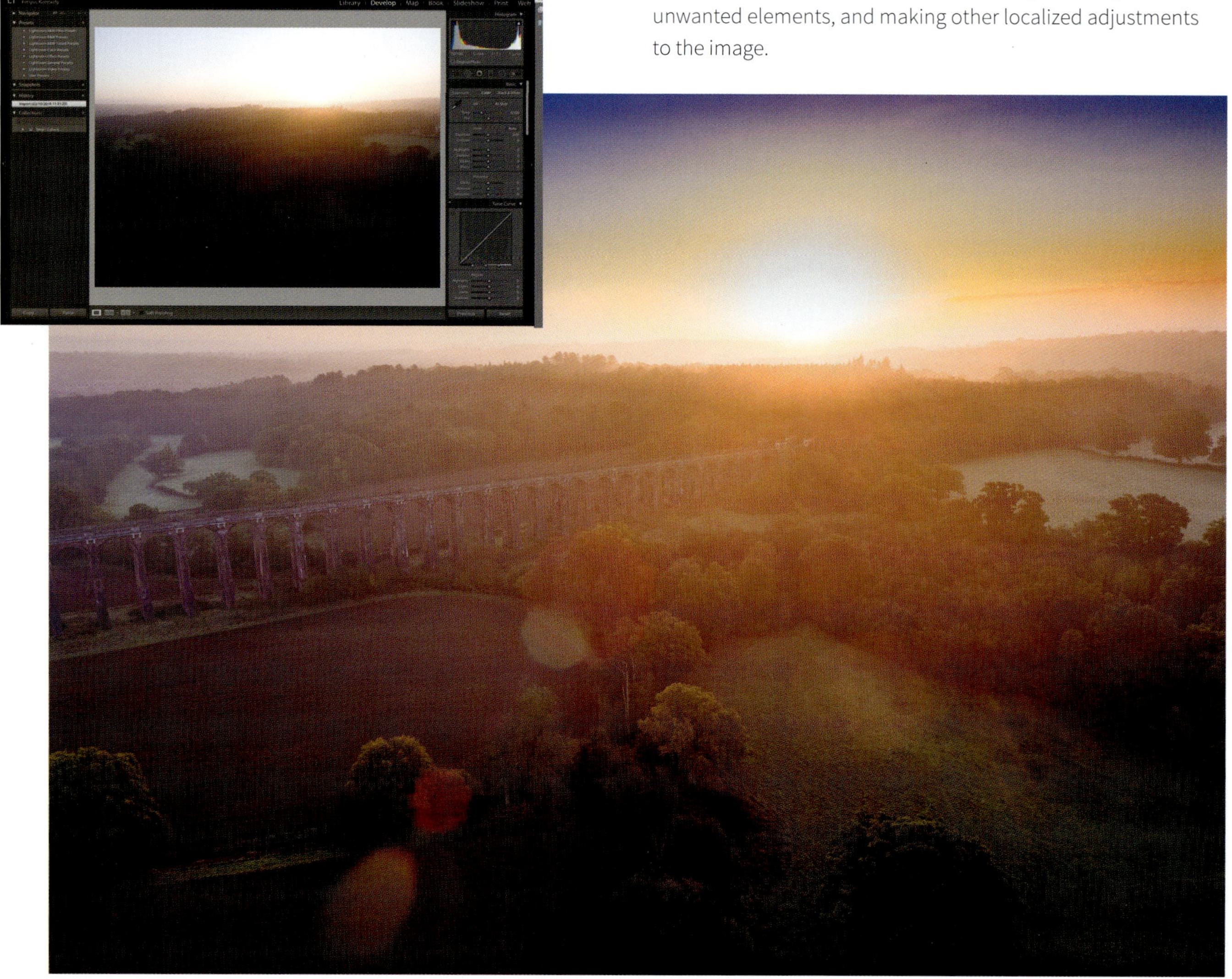

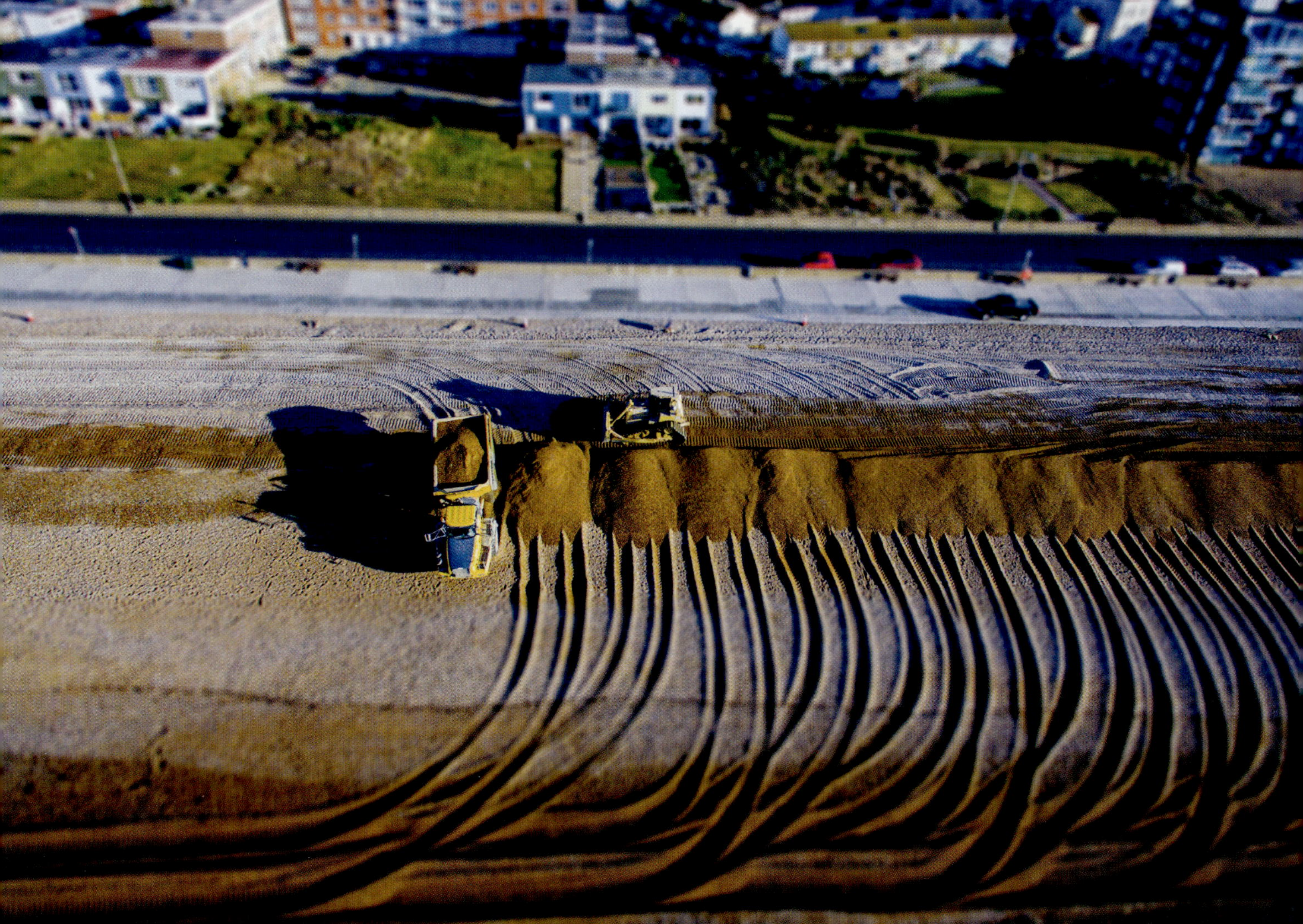

▲ Blurring everything apart from a narrow strip of the image gives the illusion of a miniature world. This effect—known as a "tilt-shift" or "miniature" effect—can be achieved either with a special lens or with software in post-processing. Some cameras also offer it as a "creative" shooting option.

◀ Shooting in Raw format gives a lot more flexibility for making adjustments with software such as Lightroom. Here you can see the unedited Raw image and the final processed shot.

TILT-SHIFT EFFECT

A "tilt-shift" lens is a specialist lens that allows the photographer to manipulate the lens plane relative to the image plane or shift it parallel to the lens plane. A popular use for a tilt-shift lens is to create a "miniature" effect. This is achieved by minimizing the depth of field so there is only a very narrow strip of focus, which creates the illusion that the subject is tiny. This technique works particularly well with elevated shots (as it enhances the idea that we're looking down on a model), so is well suited to drone work.

Unfortunately, a drone large enough to lift a DSLR with a tilt-shift lens is pretty expensive (as are the lenses themselves). Fortunately, it's possible to "fake" a tilt-shift effect in post-processing, either by applying a pair of gradient blurs or by using a dedicated tilt-shift filter in your editing software.

HIGH DYNAMIC RANGE

Sometimes you will find that you have a very wide brightness range in your image and that if you expose correctly for the sky, the land will be very dark (and the sky will burn out if you expose for the land). In this situation you can try photographing the same scene at different exposure settings, so you have a series of images that run from light to dark. This sequence can then be blended in specialist software to make a single high dynamic range (HDR) photograph that uses the correctly exposed parts of each shot.

The main thing here is that the content in each image should be identical, apart from the overall exposure. To start with, you should try to ensure that the camera doesn't move between shots, and that subject movement is kept to a minimum. HDR software can often compensate for slight movement between frames, but severe misalignment can result in unwanted artifacts in the final image.

You should also try and "lock down" as many of the camera settings as possible. This means setting a specific white balance, rather than relying on Auto white balance, for example, and perhaps even going as far as focusing manually to prevent the lens refocusing between shots.

Once you have taken your exposures you will need to combine them. Some image-editing programs (such as Photoshop, for example) have HDR tools built in to them, or you can use specialist HDR programs such as Photomatix Pro. In each case, the software will use "tonemapping" algorithms to combine the "best" bits of each shot and deliver your final image. However, tonemapping is an art that needs plenty of practice—it is quite easy to overdo the processing of an HDR image and end up with something that looks very artificial.

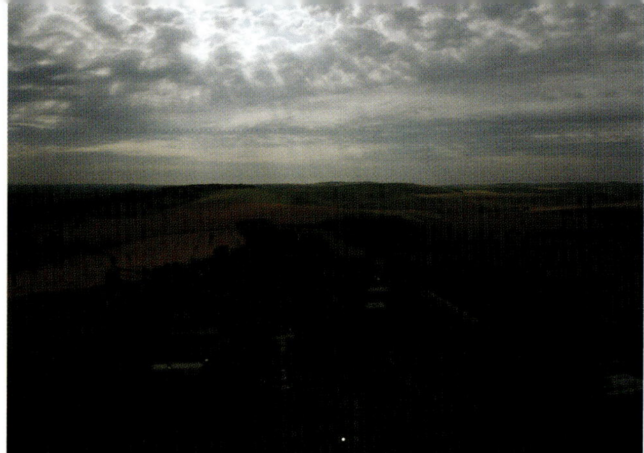

-2 stops (exposed for highlights)

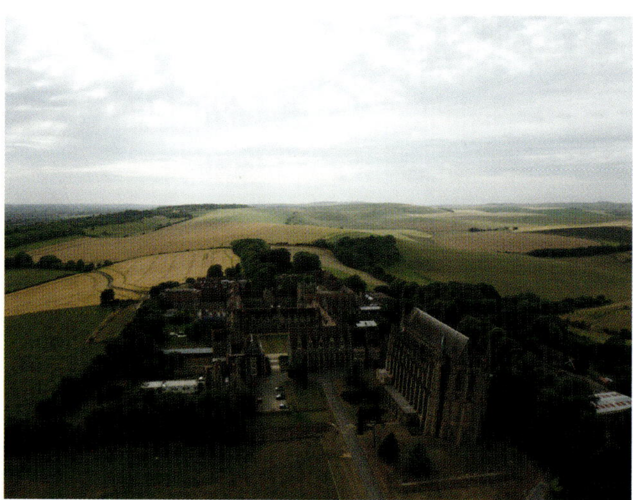

As metered (exposed for midtones)

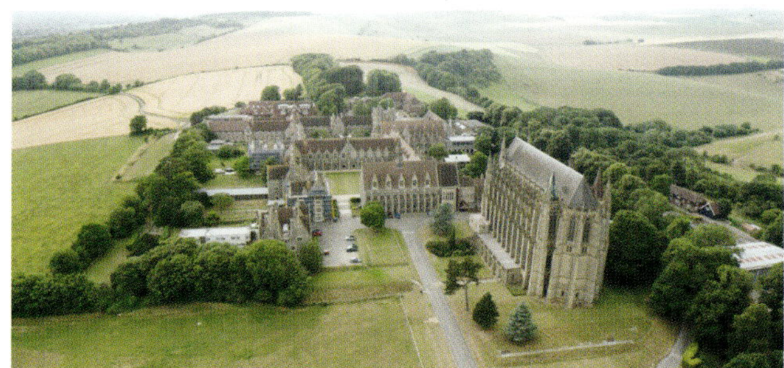

+2 stops (exposed for shadows)

▶ To keep detail in the sky and land I shot three images, 2-stops apart (right) and then combined them to create the single HDR image shown opposite.

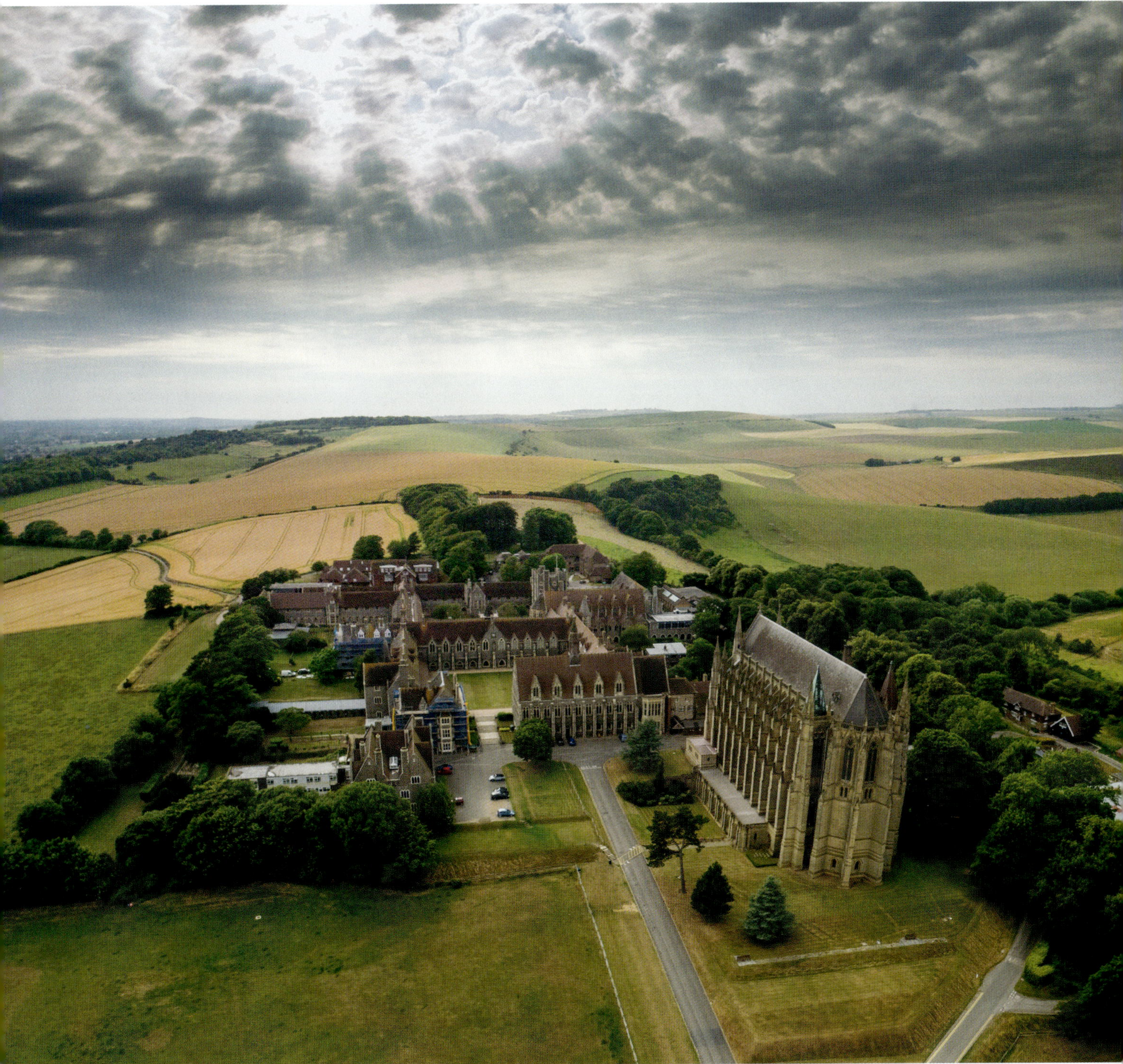

PANORAMAS

There are times when an aerial photographer needs to record a view that's wider than that allowed by his widest lens. This is where the art of stitching multiple photographs together can come in very useful. The underlying principle is that you shoot several photographs that overlap and use your editing software to merge them seamlessly into a single image covering a much wider area. This is often not a particularly difficult task, but there are a few things to be aware of.

LOCK IT DOWN!

One of the most useful features of a modern drone for the panorama shooter is GPS stabilization. This enables the pilot to maintain a hovering position without any input to the controls. This is crucial for panoramas, as any movement of the camera between shots can introduce troublesome parallax errors that make stitching a difficult or impossible task. GPS position hold is therefore the equivalent of using a tripod at ground level.

With a stitched panoramic image the camera should ideally rotate around its nodal point (the point within the lens where the light rays converge). However, the rotation of the drone or gimbal on its own center of rotation ("yawing") is sufficiently accurate for most aerial panoramas.

As well as the drone's physical location, the camera settings should also be locked. If there is a large variation in light across the area you want to photograph, a camera set to Auto will change the exposure for each shot in the panorama. This can be problematic for some stitching software and lead to uneven results. It is far better to choose the most critical part of the image and expose correctly for that. The settings should then be "locked in" for all the images in the sequence. Focus and white balance should also be locked if possible.

If your camera/drone setup does not allow you to set the exposure manually, don't despair! It's still possible to achieve very successful panoramas, it's just that your software may have to work a little harder after the event.

▼ Even with the drone at an altitude of 400ft (120m) this pier did not fit into a single frame. Instead, 15 shots were stitched together with Adobe Lightroom.

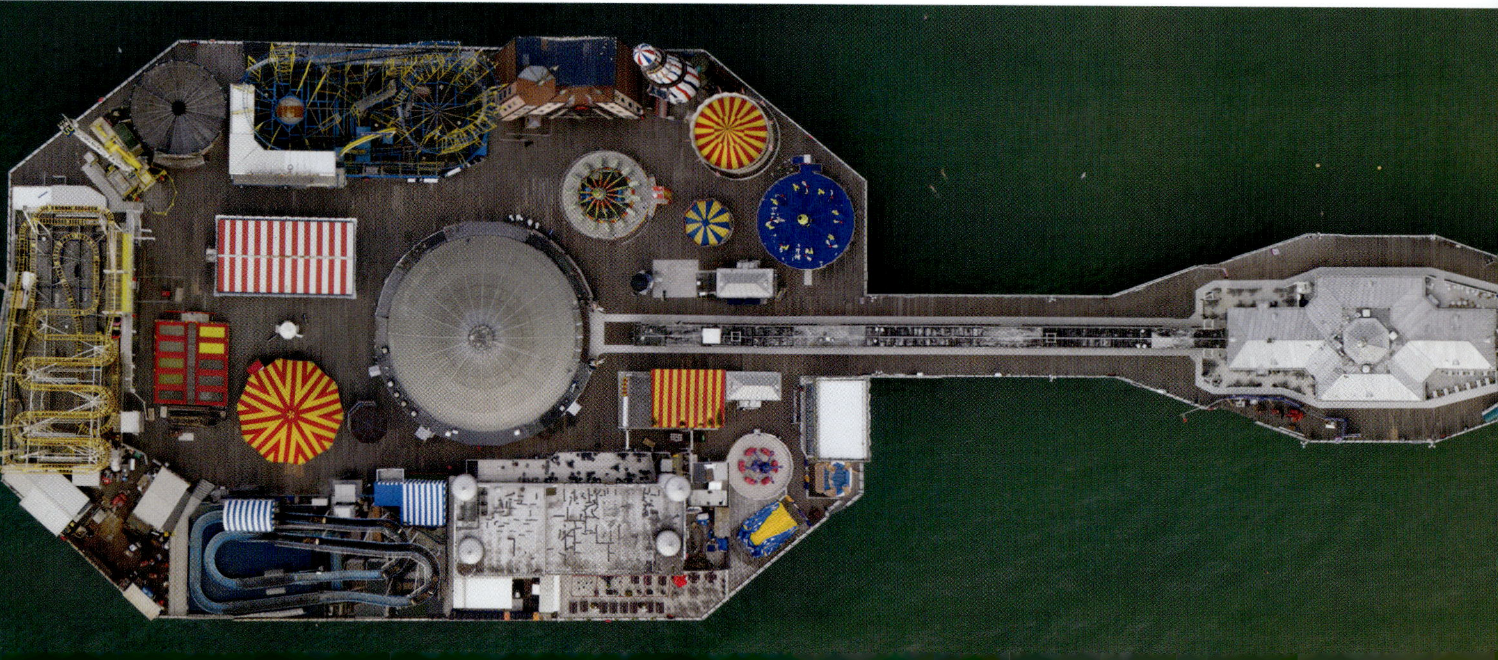

SHOOTING YOUR FIRST AERIAL PANORAMA

Let's imagine you've run through all of your preflight safety checks and planning, you've envisioned your panorama, and you're ready to shoot. Depending on your drone setup and whether you are working as a one-man or two-man operation, you will either be yawing your whole drone or panning the camera to cover your area of interest.

Before you shoot a frame it's worth giving some thought to the composition. How far will your panorama go? Is it a full 360 degrees, or just a section of the scene that is of interest? Start by identifying the beginning and end frames by yawing your drone.

When everything is ready and you've double-checked your camera settings, frame the first shot and press the shutter-release button. Then, taking care not to touch any flight controls other than the yaw, turn the aircraft carefully (I usually turn from left to right by habit) until you have an overlap of around 30–50% with the previous image. It helps to use a "landmark" in the center of the previous image and move it to somewhere close to the edge of the next frame. This will ensure plenty of overlap for the stitching software to work with.

When you're happy with the overlap, press the shutter-release button and repeat the process until you reach your final frame. A full panorama can consist of anything from two frames to many hundreds if you're shooting an extremely high-resolution, 360-degree panorama with a long lens.

LENS CHOICE

The wider the lens you use, the fewer shots you will need to complete the panorama, so a fisheye is probably the best choice if it's available to you. If not, a wide-angle rectilinear lens—such as those found on the DJI Phantom range—will also do a good job.

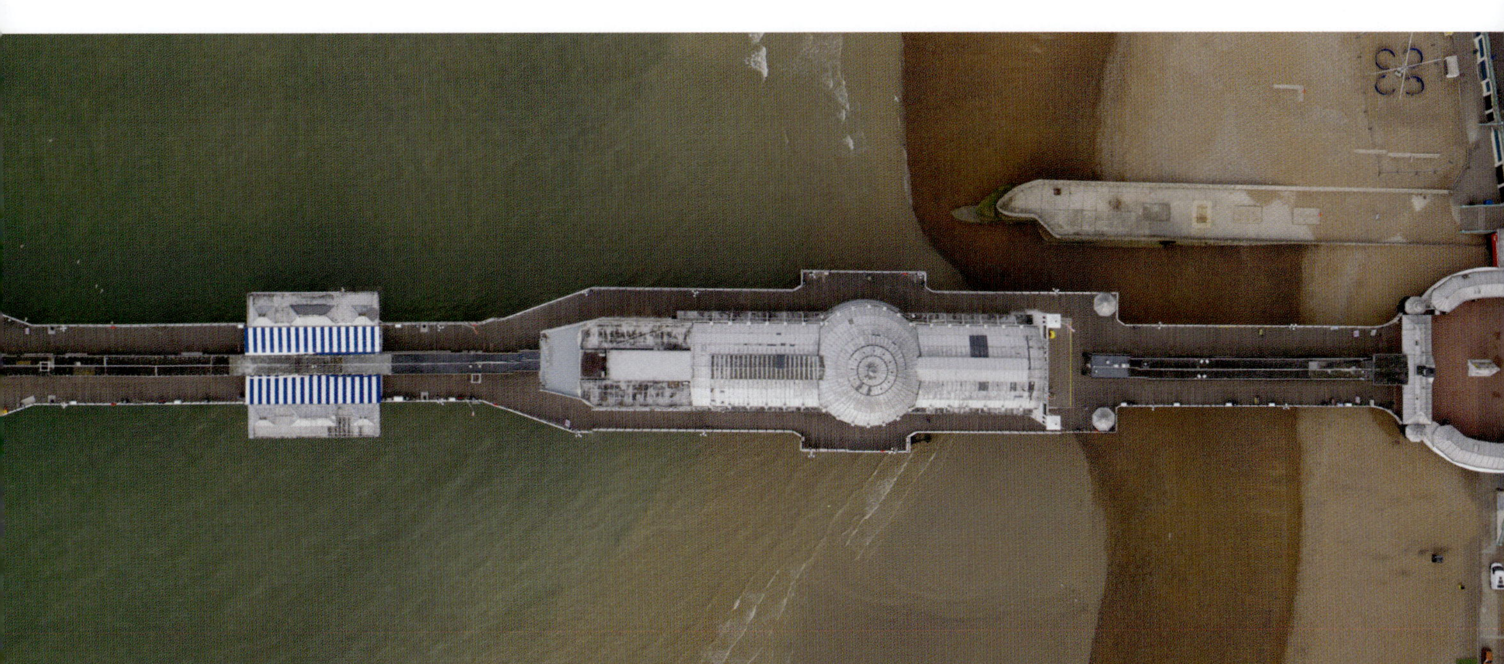

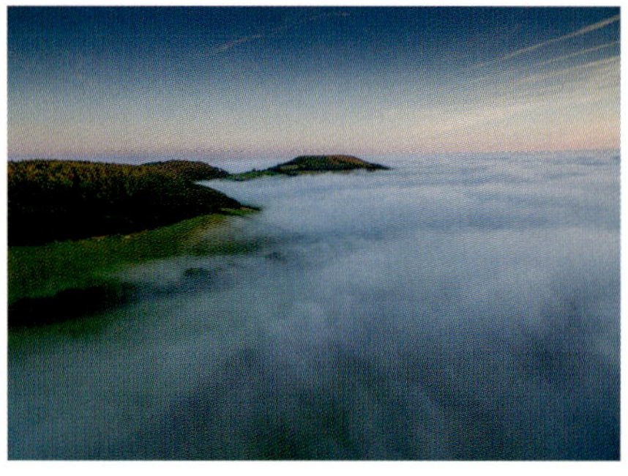
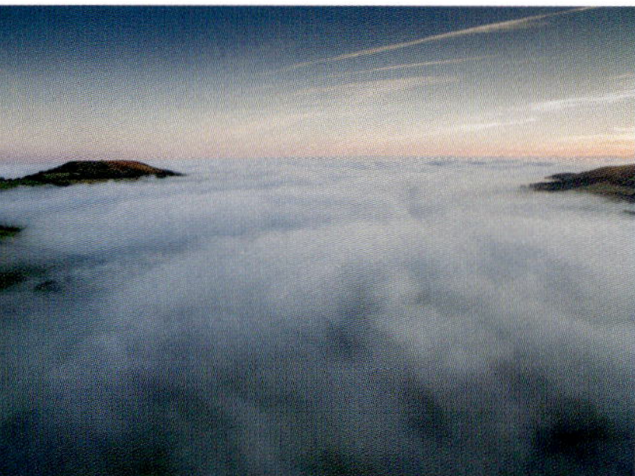
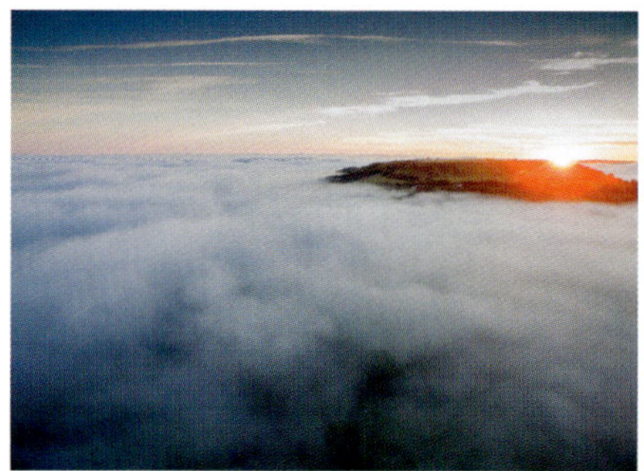

MULTI-ROW PANORAMAS

Once you have mastered single-row panoramas you can move on to multi-row panoramas, which can potentially cover the entire sphere of view. These "spherical panoramas" can be used for a variety of applications, although the basic output is an "equi-rectangular" panorama that covers 360-degrees horizontally and 180-degrees vertically.

In practice it is usually either not possible or not desirable to include all of the upward view of the sky (and the drone itself), but this can be "cheated" as I will explain later. The equi-rectangular panorama can then be output in a variety of formats, including as an interactive virtual reality panorama, or as a "little planet" (technically known as a stereographic projection).

OPTIMIZING THE PANORAMA

Large areas of water or sky in a panorama can cause stitching problems, as they're often either featureless (a plain blue sky, for example) or mobile (waves and ripples on water, or clouds scudding across the sky). You should therefore try to plan your panorama so that each frame has enough static features in common with neighboring frames to allow accurate stitching.

You may also find the stitching software struggles if there are a lot of moving objects in a scene, such as people or cars. However, these problems may be rectifiable with a spot of manual manipulation in your software of choice.

A multi-row panorama takes the idea of a standard stitched panorama and builds on it by adding multiple rows of images, both above and below the horizon. These are shot by tilting the camera vertically, again maintaining a 30–50% overlap (not only with the images that are adjacent horizontally, but also those that are adjacent vertically). This process is continued until the full extent of the terrain below the drone is covered.

It is usually enough to shoot only one extra row of shots above the horizon, and it will help if the horizon is included in all the shots (toward the bottom of the frame), as the software will struggle to find points to match up in a relatively featureless sky.

Most camera/gimbal combinations do not allow you to angle the camera directly upward, and even if you could, the drone itself would dominate your view. The best solution is usually a slight "cheat." After you've landed—and if it's feasible—remove the camera from the drone and hand shoot some additional shots to cover the missing patch of sky. These can be used to patch the hole where the drone was.

▲ This panorama (above right) is the result of shooting and stitching six images together (top). A small yaw of the drone was made between each shot, covering more than 180 degrees. To ensure consistency the exposure settings were locked, while Raw capture guaranteed high quality.

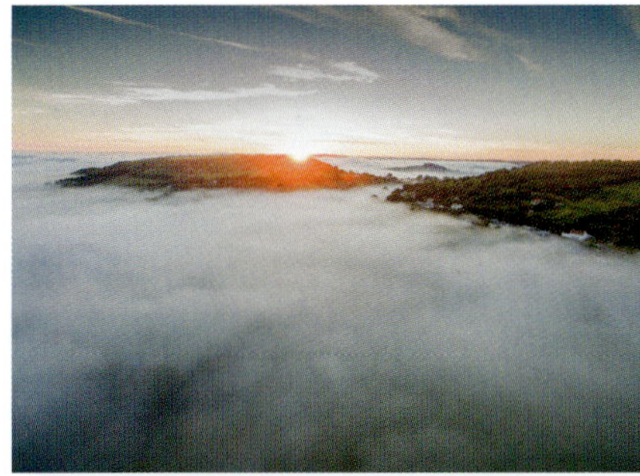

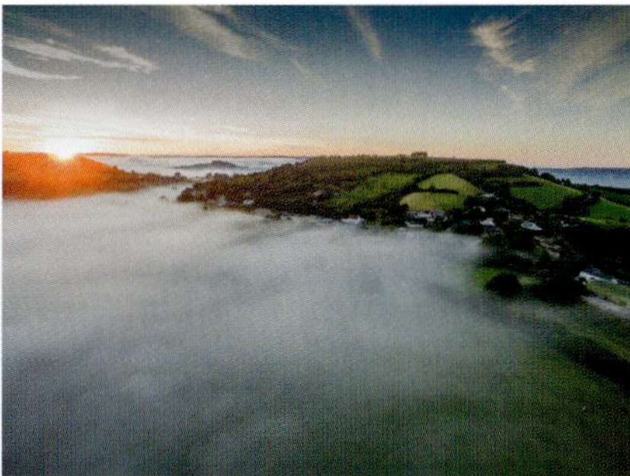

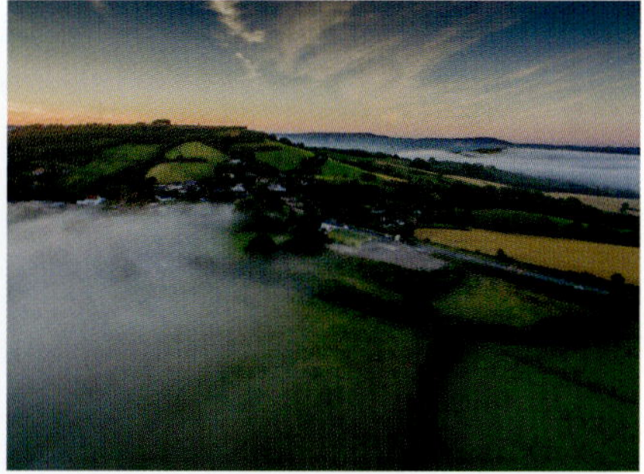

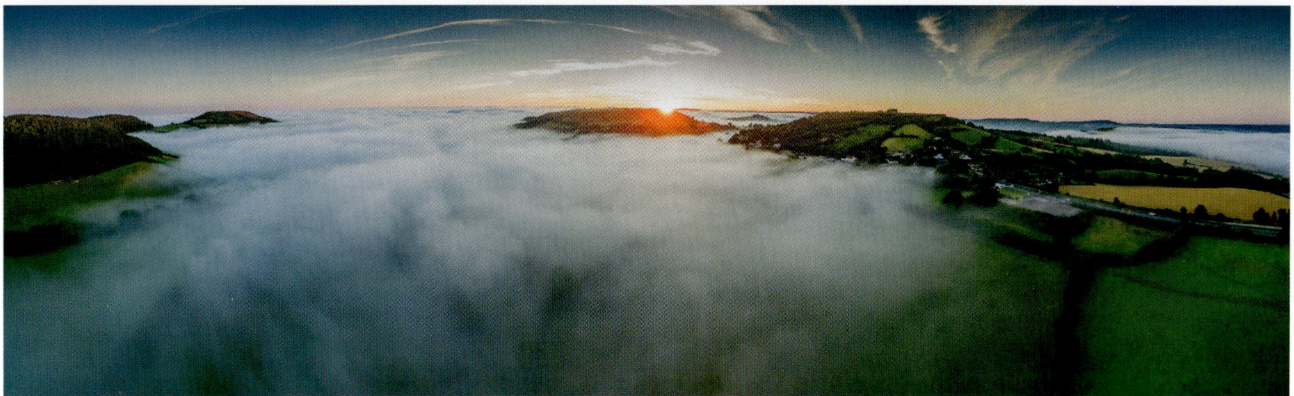

GOING AUTOMATIC

Increasingly, there are now methods that automate the panorama process. In the case of the DJI Zenmuse 5D Mark III gimbal, this involves firmware built into the gimbal. A single switch on the remote can be programmed to initiate a sequence of gimbal-rotation and shutter-triggering actions, resulting in around 30 overlapping still photographs covering most of the sphere of view.

For those with a DJI Phantom 3 or 4 model—or the DJI Inspire—there are third-party phone apps that include an automatic panorama function in which you can specify the number of columns and rows of photographs in the panorama for the drone to capture.

PROCESSING

Back at your computer, your set of perfectly exposed and neatly overlapping images will be fed into software that will analyze them, identify "control points" that are common to each pair of overlapping images, and (with varying degrees of success) warp and massage neighboring images so they blend seamlessly with one another.

Popular programs for this task include Adobe Photoshop and Adobe Lightroom (which I've found often does a very good job with single-row panoramas). For those who want more control over the stitching process there are dedicated programs, such as PTGUI Pro, that tend to offer more advanced output options.

PLANNING SHOTS

Planning your shots in detail will reap rewards, and as with your safety clearances, you can often get some idea of the lay of the land using Google Earth. The more time you spend planning, the higher chance you have of being successful in your aims, so after you have considered issues of safety, weather, and air traffic, you should invest some time in planning your photographs, with the following in mind:

Lighting: Think about the ideal lighting for your shot. Where should the sun be? Sometimes this is a simple question—if you are shooting still shots of a house for a realtor or estate agent, for example, you need to know which is the best side of the house and at what time it will have sun on it. More complex shots might require the sun to exactly line up with a particular landmark at sunset for instance. There are plenty of great tools that can give you the exact angle of the sun (or moon) on a map at any time and date, including *The Photographer's Ephemeris*.

Tide times: If you are planning a coastal shot, be sure to check the tide times, both from a safety and access point of view, and an esthetic point of view. Often a particular tide state is critical for a composition—low tide might be needed to expose some beautiful patterns on a tide flat, or you might want to shoot at high tide so the sea is splashing directly onto the bottom of some cliffs. Tide tables for your location can usually be viewed easily after a simple web search.

Height: How high will you need to fly to get the shot you want? For top-down shots or mapping applications it's often very important to know how high you should fly to get the composition you need. You should be able to work this out knowing the field of view of your lens (found on the camera or lens' spec sheet) and the dimensions of the area you need to shoot (which you can get from Google Earth).

You can then use some basic trigonometry or a phone app such as UAV Coverage to help. This will tell you if you can get the planned shot from lower than 400ft (120m). If not, all may not be lost, as you can take multiple overlapping shots for a stitched higher resolution panorama. If you have interchangeable lenses, you may also be able to switch to a wider-angle lens for the job.

Positioning: Give some thought as to the best place to stand for a good line of sight to your drone. This is particularly important if you are shooting video and need to fly close to obstacles, as you need to be able to judge your distance and your flight line. Generally, the closer you are, the better.

Clothing: As well as wearing clothing appropriate to the weather, give some thought to the clothing of anyone featured in your drone shots. Human subjects are often fairly small in the frame in an aerial shot, so getting them to wear clothes that will stand out against the background can be very important.

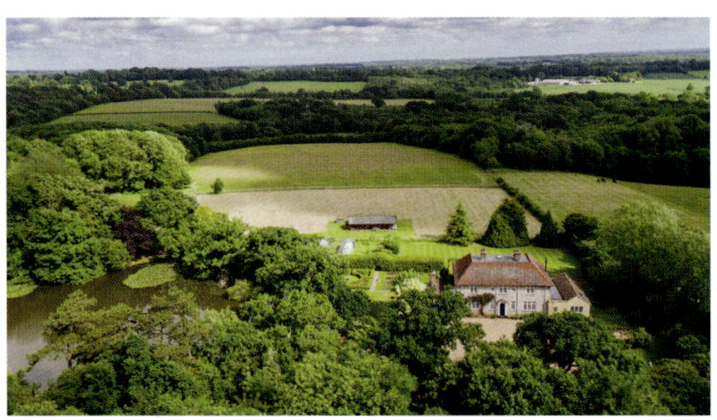

◀ Time of day has a big influence on how your shot looks. If you're photographing a house for instance, you need to make sure the sun is hitting its best side.

▶ Lighting (and the shadows it created), the height and position of the camera, the tidal state, and the subject's clothing and accessories all played a major part in the success of this top-down coastal shot.

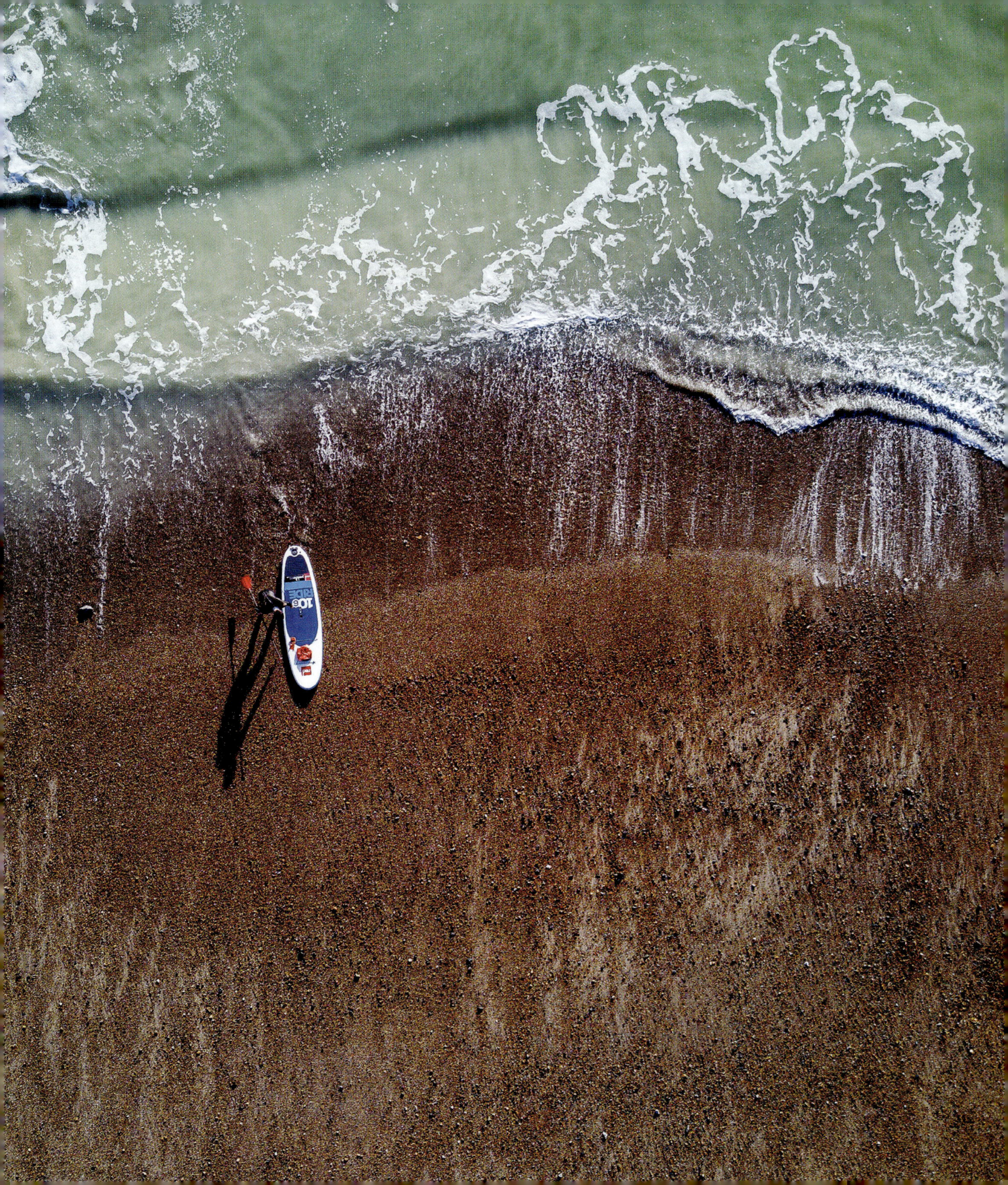

PROFILE: DAVID HOPLEY

David Hopley is a self-taught, award-winning photographer with a background in architecture and graphic design, which has given him a keen eye for detail and composition. Having quickly spotted the potential that aerial photography has for producing unusual and arresting compositions, David bought a drone and started creating amazing and often abstract images.

"Drone photography has literally given me a world of new opportunities when it comes to creating unique landscape images and it's completely changed my approach to landscape photography. Quite often it is impossible to visualize a 'top-down' shot until I am flying above it; there are hidden textures and patterns in the landscape that just cannot be seen at ground level. You should never assume that because a scene or object does not look great at ground level it will look equally poor from the air—don't forget that most man-made structures and landscapes are designed and drawn in plan form, so they often have pleasing esthetic features or have strong geometric shapes and patterns when you look down on them.

More often than not, I will head out to a location with no idea what I will be photographing. Once I have launched the drone, I will quickly scout out the location, looking for features that could make an interesting image. The drone gives me the freedom to search the landscape in a way that might be inaccessible by other means, and a very wide area can be explored during a 15-minute flight. So no matter what you seen from ground level, power up the drone and go and take a look at it!"

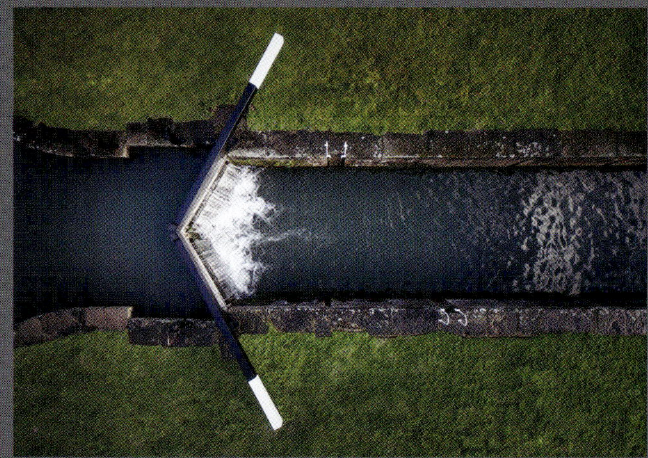

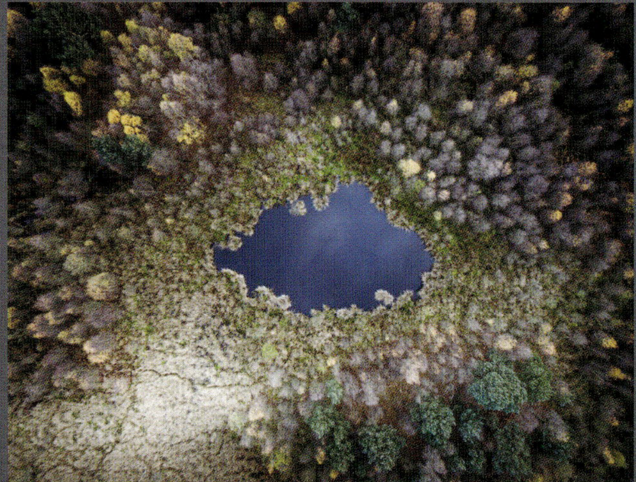

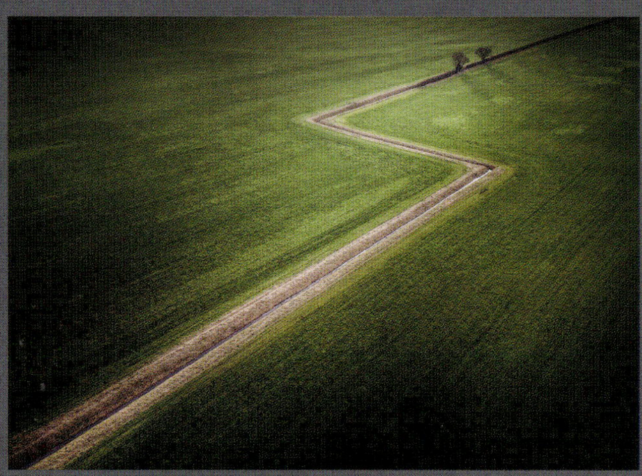

▶ David's design background is clearly evident in his often abstract aerial images, which typically display a strong, graphic use of color and composition.

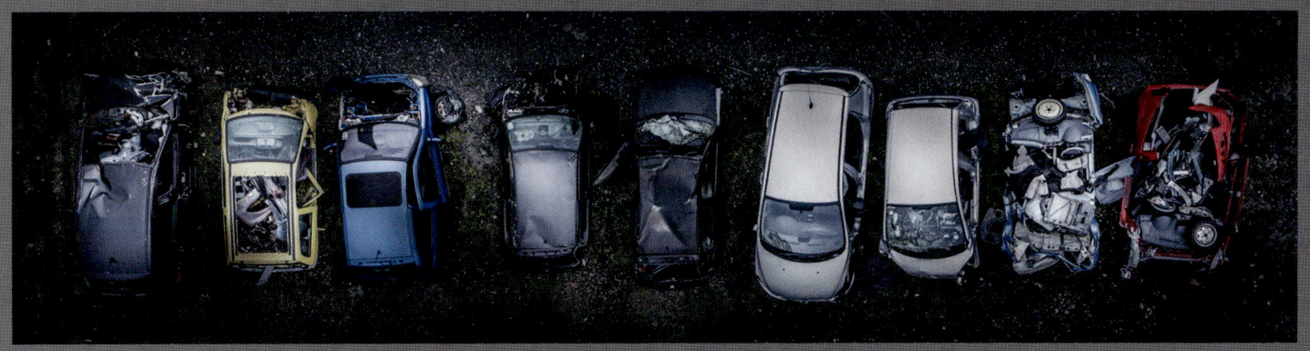

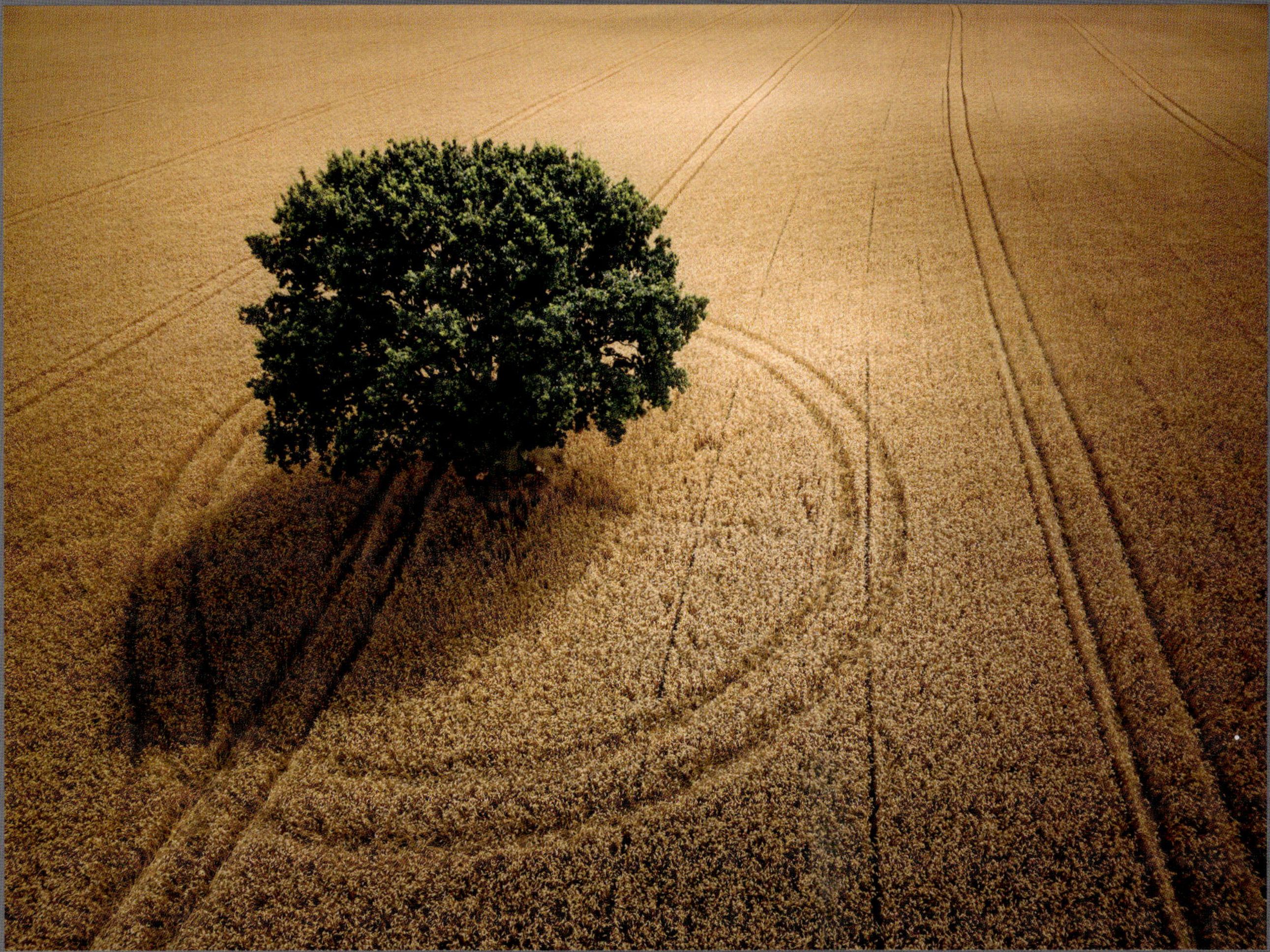

CHAPTER 06

SHOOTING VIDEO

To an experienced still photographer, shooting video might at first seem like a simple task—after all, it's just a series of still frames strung together. However, as a quick visit to the set of a major movie or TV production will soon tell you, it can be a very complicated process that often requires the involvement of large numbers of people (an indicator that there's too much for one brain to process!).

This chapter deals not only with the technical details of how to set up a camera and drone for shooting video, but also with how to execute a video shot so that it adds to your story, regardless of whether it's a shot of your family in the garden for a home video, or a crucial action scene in a Hollywood blockbuster.

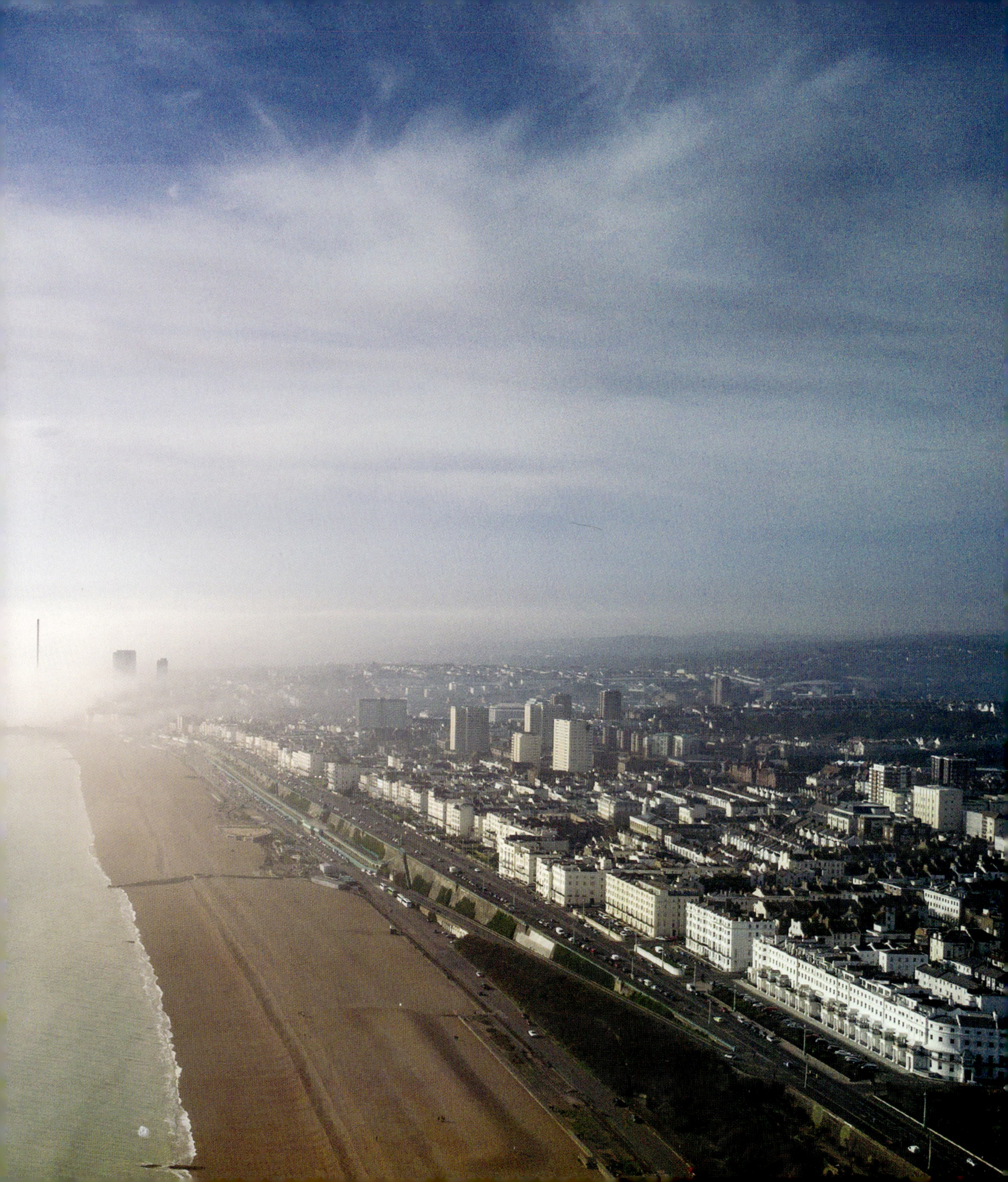

ONE PERSON, TWO JOBS

While a still photograph involves careful framing, followed by a momentary shutter press, a video is a protracted event lasting from less than one second to several minutes. As a result, shooting video from a drone requires you to concentrate more heavily on both flying the drone *and* operating the camera.

As a drone pilot is ultimately responsible for the safe operation of the aircraft, this must always be the priority. Relatively simple video shots in wide-open environments can be straightforward for a novice pilot. Such shots might include a vertical rising shot or forward movement over fields or the coastline—essentially situations where the pilot can safely look at the screen without fear of colliding with obstacles.

More complex camera moves—where the drone is moving on several axes—can be achieved by more experienced pilots operating as a single operator, but in a complicated environment it is usually much safer to separate the tasks of pilot and camera operator and have two operators: one working the camera and the other flying the drone.

▼ Separating the roles of pilot and camera operator enables both people to concentrate solely on their part of the process, which helps ensure the safest possible flight *and* the optimum video footage.

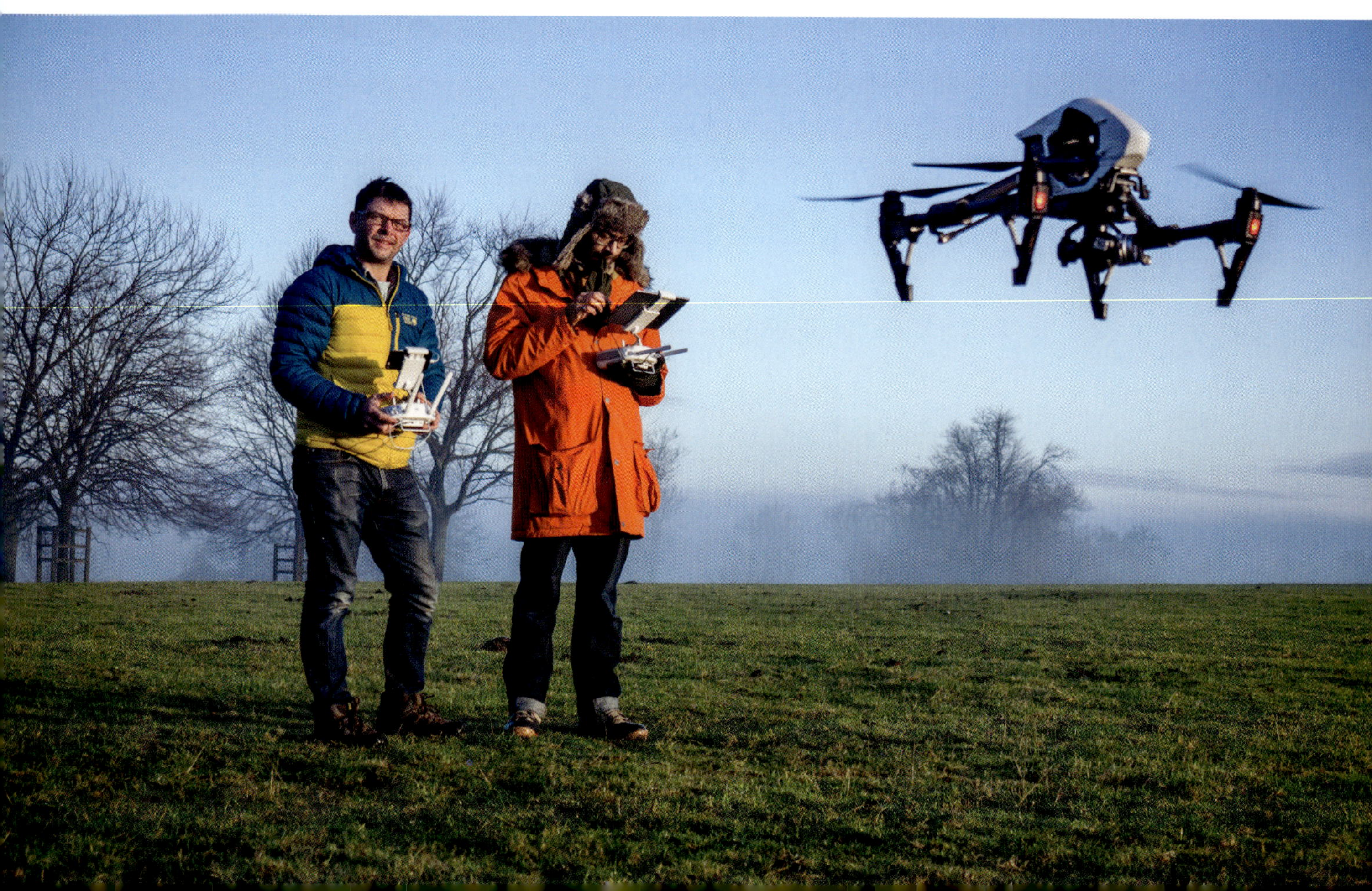

WORKING TOGETHER

A good pilot/camera operator team should be in constant communication with one another. Ideally, the camera operator will take care of all camera-related tasks, from the preflight shot planning and camera setup, through to the execution of the shots (including exposure, focus, and so on). This frees up the pilot to concentrate on preflight aircraft safety checks and flying the drone.

In flight, the movement of the drone will need to be coordinated with the movement of the camera. Typically, the camera operator will suggest the shots beforehand and discuss them in detail with the pilot. He or she may also be the main point of contact for any client or director who is at the location, taking in any suggestions and—if necessary—relaying them to the pilot at an opportune moment.

GIMBAL ISSUES

Symptoms of a badly balanced or tuned gimbal may include sudden shakes or wobbles in your footage as the motors struggle to cope with the demands placed on them. Continuous high-frequency vibration of the gimbal may also result in an effect known as rolling shutter or "jello," whereby the entire image seems to wobble like jelly. This happens because the camera reads data from the image sensor one line at a time, and although this happens quickly, any movement in the image during this time can lead to the lines being out of register when they are read.

The vibrations that cause jello can be the result of a defective or poorly set up gimbal, or due to an unbalanced propeller. Mild jello effects can often be overcome with the use of slower shutter speeds and ND filters (see page 142).

GIMBAL SKILLS

More than any other piece of kit, it is the electronic gimbal that has enabled smooth, high-quality video to be shot from relatively small multirotor aircraft. Not only does a good gimbal eliminate unwanted shakes and vibration from the aircraft, but it allows the operator to aim the camera in the desired direction.

On drones with integrated cameras and gimbals, the user does not need to worry too much about setting up the gimbal correctly. However, there are settings—often found in the drone's app—that will allow you to fine tune the response speed of the gimbal. Although panning of the camera will be the control of the aircraft yaw, tilting the camera is often controlled via a wheel on the remote control unit. By default, this often leads to a fast and jerky movement, but slowing the response speed can give a smoother, more cinematic look to the camera movement.

With a two-man operation team, the range of gimbal control options available can grow. For example, sticks can be assigned to the different axes of movement (tilt, pan, and roll); to horizon level correction; and the speed of movement of the gimbal on each of these axes. The options you use will depend on the particular shots you need and the stick assignment the camera operator feels most comfortable with.

With larger cameras on larger drones, the setup of the gimbal becomes progressively more complicated, especially when it comes to generic gimbals that are intended to work with a variety of different camera and lens combinations. It is not only critical that the gimbal is balanced correctly (see page 37), but it is also often necessary to fine tune the power of the gimbal on each axis to correspond to the camera/lens weight. This whole process can easily take an hour or two, but the good news is that once it's done for a particular camera/lens combo you should be able to leave it alone.

GET MOVING!

Camera motion can add an incredible sense of dynamism to a video shot. But it has to be executed correctly and at an appropriate speed for the purpose. It's a common mistake to move too fast, often a small amount of movement is all that's needed. The options for combinations of drone and camera move are almost limitless, but here I'll outline a few basics.

STATIC SHOTS

The most basic of video shot is a static shot, where the drone is flown to a chosen spot and left to hover on GPS lock. This is a "locked off" shot, which gives you plenty of time to compose the shot nicely. In many ways, it's similar to using a conventional tripod on the ground, with its head "locked down."

If there is a good amount of action and subject movement, this can be an effective technique. Static shots also work well for direct, top-down shots, which provide an unusual perspective that can take the viewer a moment or two to interpret.

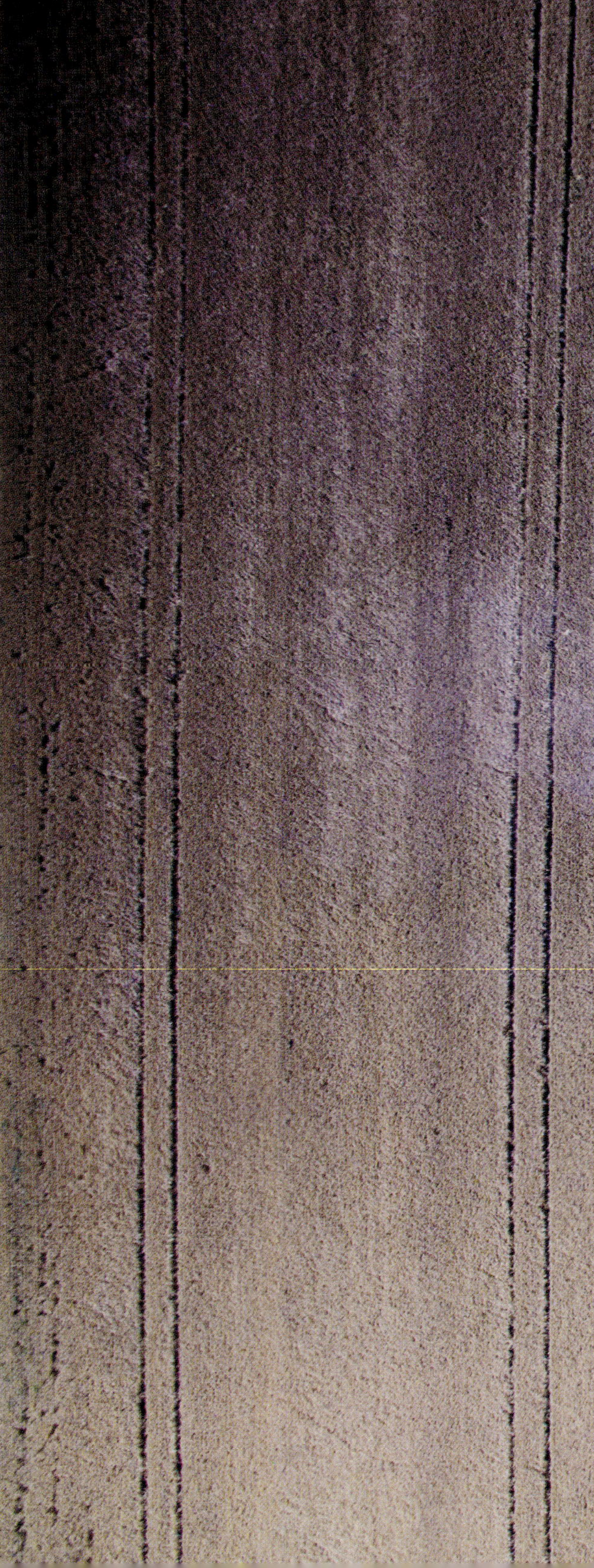

▶ Sometimes a static shot achieves the desired outcome: when the subject is moving you don't always need any additional movement from the camera.

SPEED

One factor to consider when planning a drone video shot is how fast you will be flying. This will depend on a number of factors, including (but not limited to) whether you want the subject/ground to appear to be passing by quickly or slowly; altitude and subject distance; and focal length.

If you are flying very close to your subject there will be a much greater sensation of speed. In fact, it's often difficult to fly a drone slow enough when the camera is very close to the subject or ground.

On a recent shoot, for example, a director needed a shot where the camera moved along a path toward a fountain, a couple of feet above the ground. In trying to achieve this, we ran into a couple of issues. The first was speed modulation. Due to the way the controls were set up, it was very difficult to maintain a slow, steady (walking pace) movement. This was partly rectified by adjusting the remote control parameters so a tiny movement of the stick had minimal effect on speed.

The second problem was that very small aircraft wobbles caused by GPS inaccuracies or small gusts of wind became very noticeable with the camera so close to the ground. After several frustrating attempts we found it was much easier and quicker to film the shot with a handheld gimbal, rather than using a drone!

TYPES OF MOVE

The next step from a truly static shot is to use the drone like a tripod; panning and tilting the camera either to follow a moving subject or to reveal a subject or landscape. There is also a selection of more advanced camera movements that might be useful in different situations, as outlined below. This is by no means an exhaustive list, and in many cases you can combine several movements into a single shot.

Crane up: A camera movement directly upward can be a very effective way of revealing a subject or setting, and it's also quite easy to achieve from a technical point of view. To start with you can leave the camera pointing in one direction and simply worry about moving the drone smoothly upward, but as you develop your skills you could try combining the upward movement with a slow tilt down of the camera to keep the subject framed nicely.

A crane up with the camera pointing directly downward can make an interesting shot, revealing more of the setting as the drone rises upward.

Tracking shots: A great shot can be achieved by flying at a low level on a course parallel to a moving subject, such as a runner, cyclist, or vehicle. If there are objects between the subject and camera (such as trees or buildings), this can enhance the sensation of speed, as will a longer focal length lens.

Successful tracking shots can be achieved with a single operator or dual-operator setup, but as the direction of travel is 90-degrees to the camera, it will be easier if you can use a separate camera operator. In either instance, the pilot will need good visibility of the drone in relation to obstacles.

Point of Interest: The "point of interest" (POI) shot should be part of every drone videographer's repertoire. With this shot, the drone orbits an object or person on a circular path while the camera pans to keep the subject at the center of the frame. (Imagine a hero shot of a person standing on the summit of a mountain, while the camera circles them.)

A single operator can achieve a good POI shot flying manually, but it takes plenty of practice to simultaneously orbit and keep the subject in the center of the frame. Essentially it involves flying the drone to its left or right (with the elevator control), while simultaneously ruddering in the opposite direction.

With a dual operator setup, the pilot merely has to concentrate on flying a circle around the subject at the desired distance, while the camera operator concentrates on keeping the subject centered.

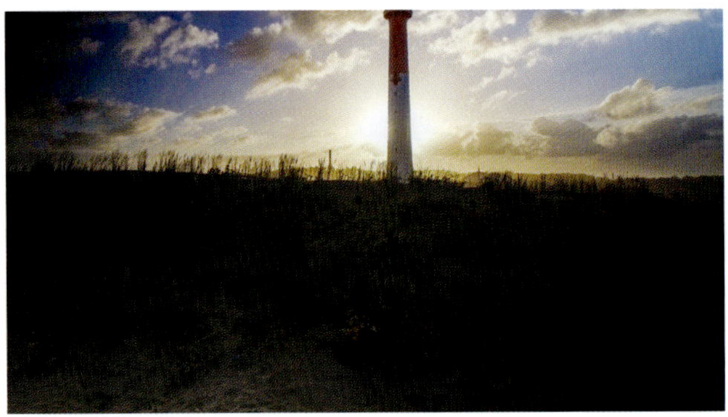

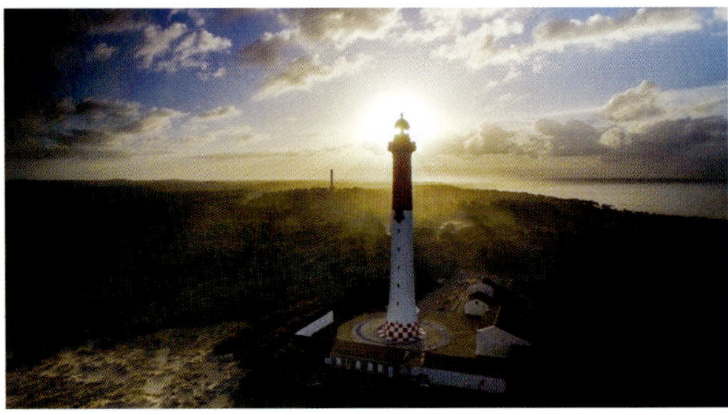

▼ A simple vertical movement, with or without a small tilt of the camera is relatively easy to achieve. This is typically referred to as a crane shot.

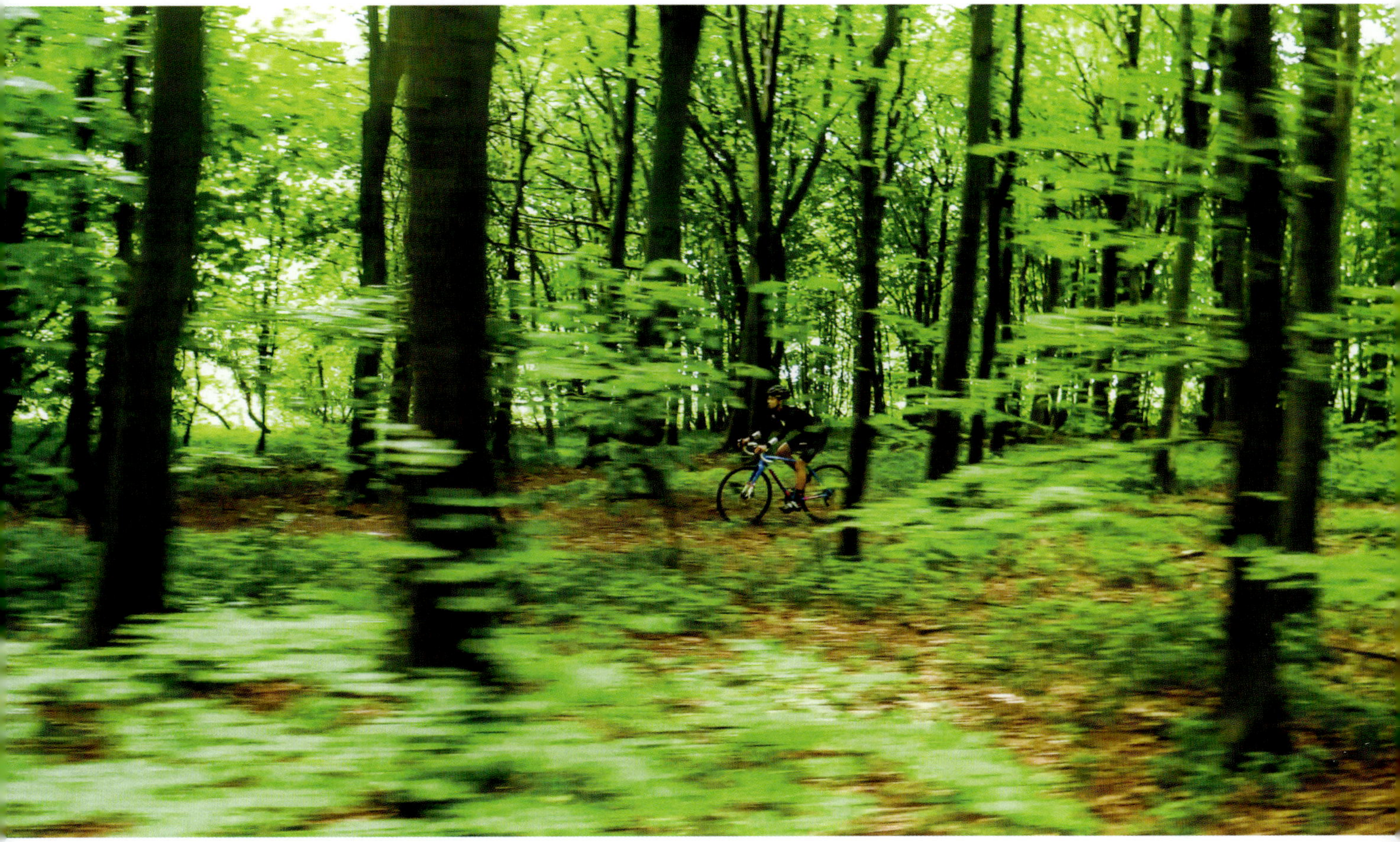

▼ A circular movement around a subject, whilst keeping it centered in the frame can look great and give a feel for the setting. This can be hard to achieve manually, but can be automated in many newer drones.

▲ In a typical tracking shot the camera moves parallel to a moving subject. This gives a great sensation of speed.

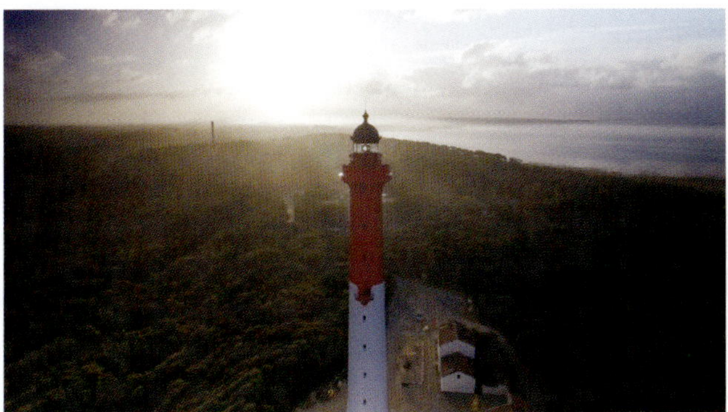

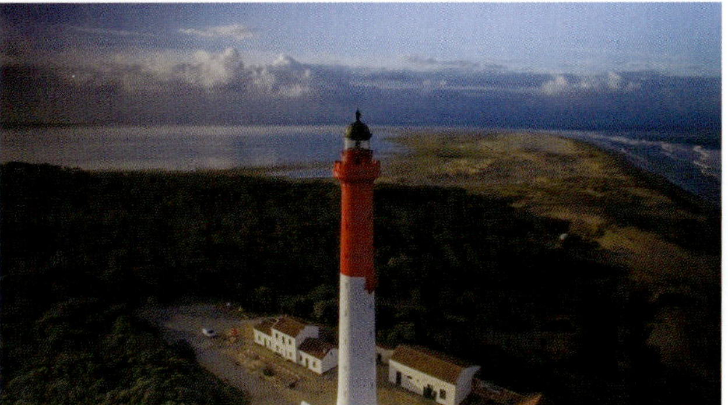

AUTOMATE IT!

Many modern drones have smart functions built into the controller app that can help you automate your camera movements in a number of ways.

Point of Interest (POI): With this feature, the drone is flown directly above the point of interest and that point is marked. The drone is then flown a chosen distance from the subject to frame the shot. The pilot can then initiate an automated orbital flight, selecting the speed and direction of orbit. During the flight, the camera will remain aimed directly at the point of interest while the drone circles it.

Automated routes: This option enables you to program a route of waypoints that the drone will automatically navigate between. This is usually done by flying manually and marking the points with the app, but once the path has been set the drone's controls are freed from the task of flying, and can be used to control the camera instead.

Subject tracking: Also known as "follow me," this feature lets the pilot identify a subject (for instance a runner, cyclist, or a car) by tapping the app or creating a "virtual" link to a mobile device. The drone will then follow the subject automatically, often with options for orbiting, side-on tracking, following from behind, and so on. A word of warning, though: there is a risk the drone may collide with obstacles in follow-me mode, although this depends on its obstacle-avoiding capabilities.

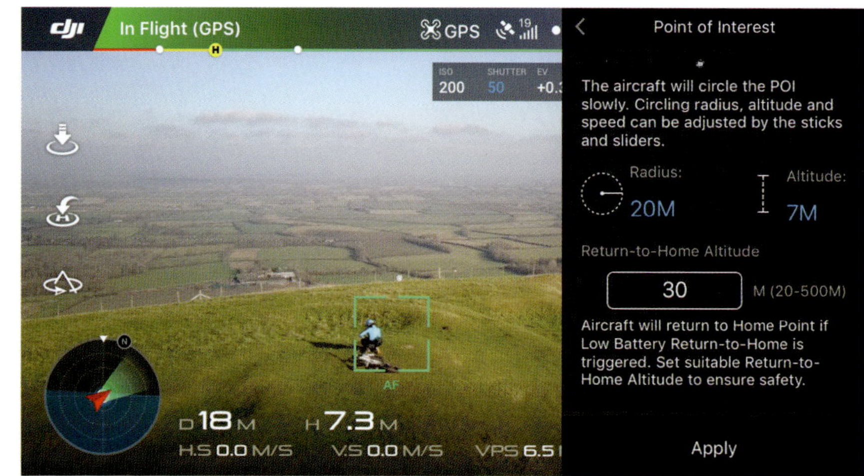

▲ Sometimes known as Point of Interest, your drone's app (DJI Go in this case) allows you to automatically orbit a subject.

▼ Clever technology in some of the newer drone systems enables them to track a subject using a visual cue. An uncluttered background and good light and contrast are essential for this technology.

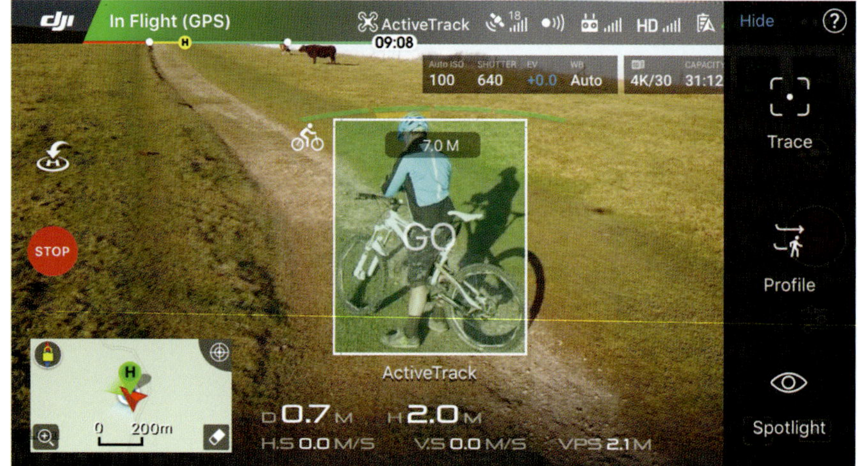

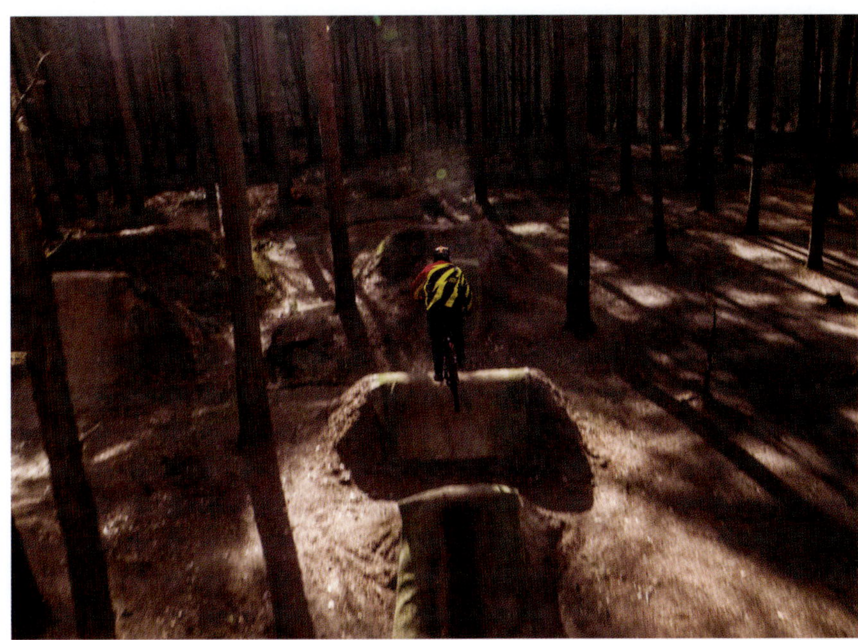

◀ When following a fast subject in an obstacle-strewn environment, it's good to have a drone with obstacle avoidance inbuilt, or preferably well-honed manual flying skills.

▼ Fast-moving subjects that change direction may outwit the capabilities of automatic subject tracking technology.

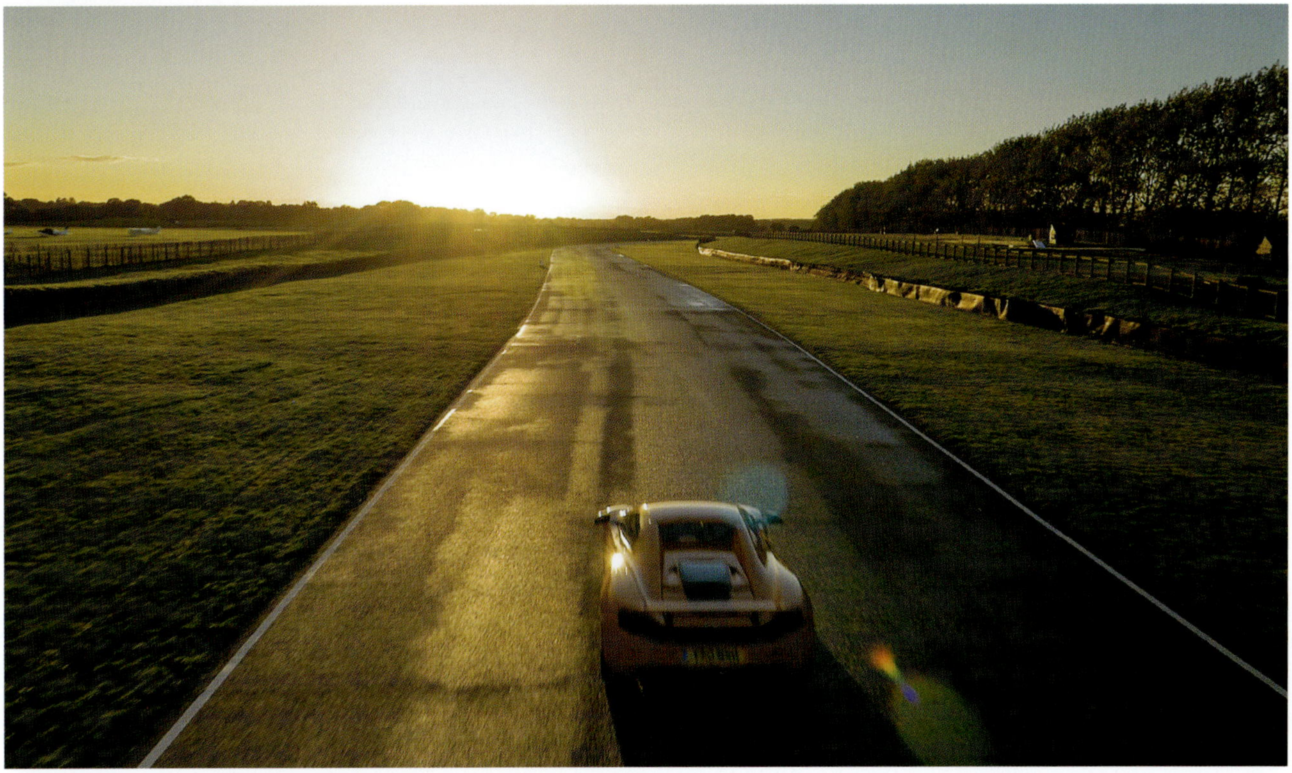

EXPOSURE & FOCUS

Getting your camera settings as close to "correct" as possible is arguably even more important when you're shooting video than it is for stills. Not only is there more to get wrong, but errors are often harder to correct after the event than they are with still images. As it's rarely one element that makes a piece of aerial footage look spectacular, there are lots of small things that you need to ensure are "just right," starting with the exposure. Wherever you can, you should lock the shutter speed, aperture, and focus so you don't experience unwanted shifts in any of these parameters during a crucial shot.

SHUTTER SPEED

When you shoot video, the rules about shutter speed differ significantly to those for still photography (as discussed on pages 108–109). The generally accepted convention—which gives a "normal" result when you watch the footage back—is derived from the days films were shot on actual film.

The "rule" is that shutter speed is set at twice the frame rate (strictly speaking *1/frame rate x 2*). If you are shooting at a frame rate of 25 frames per second (fps), for example, this means you should generally use a shutter speed of 1/50 sec.; at 30 fps the shutter speed would be 1/60 sec. This may seem quite slow, but it will make any fast movement in the frame appear smoother, as there is some motion blur of the subject between frames.

However, you might choose to depart from the "double frame rate" rule when you're shooting in very low light. Prior to dawn, after dusk, or flying indoors, you might need to use a slower shutter speed to get enough light to the sensor. Here you may go as slow as *1/frame rate x 1* (so 1/25 sec. if you're shooting at 25 fps, and 1/30 sec. at 30 fps), although there is likely be a significant amount of motion blur.

Alternatively, there may be times when you want to use a faster shutter speed. If you were to shoot at a shutter speed of 1/250 sec. or faster, for example, each individual frame will be very sharp. The downside is that any fast movement will appear "jerkier." Although this is usually undesirable, it is commonly used in "gritty" action sequences.

ND FILTERS

An irritating by-product of shooting at very fast shutter speeds can happen in bright daylight with a large aperture. If the sun is shining through the propellers and onto the lens you can get a distracting pattern of lighter and darker stripes moving up or down the footage as the prop blades' shadows pass across the lens. Shooting at a slower shutter speed will effectively blur these stripes and make them invisible, but if you can't (or don't want to) shut down the aperture this can initially seem impossible to achieve.

The solution is to fit a neutral density (ND) filter. An ND filter is a neutral (non-tinted) filter that is placed over the camera lens to reduce the amount of light reaching the sensor—like sunglasses for your camera.

Obviously, the amount of light there is varies according to the time of day, cloud cover, and so on, so ND filters are available in different strengths (measured in "stops"). You can also get variable ND filters that you turn to set the strength, although this type of filter can introduce unwanted color casts or produce variable darkness in a blue sky. The more expensive variable filters are better, but you may want to stick to a good set of fixed NDs instead.

BEWARE THE JELLO!

The jello effect (see page 135) is more noticeable at shutter speeds of 1/100 sec. and faster. Extending the shutter speed to *1/frame rate x 2* will reduce or even eliminate the problem.

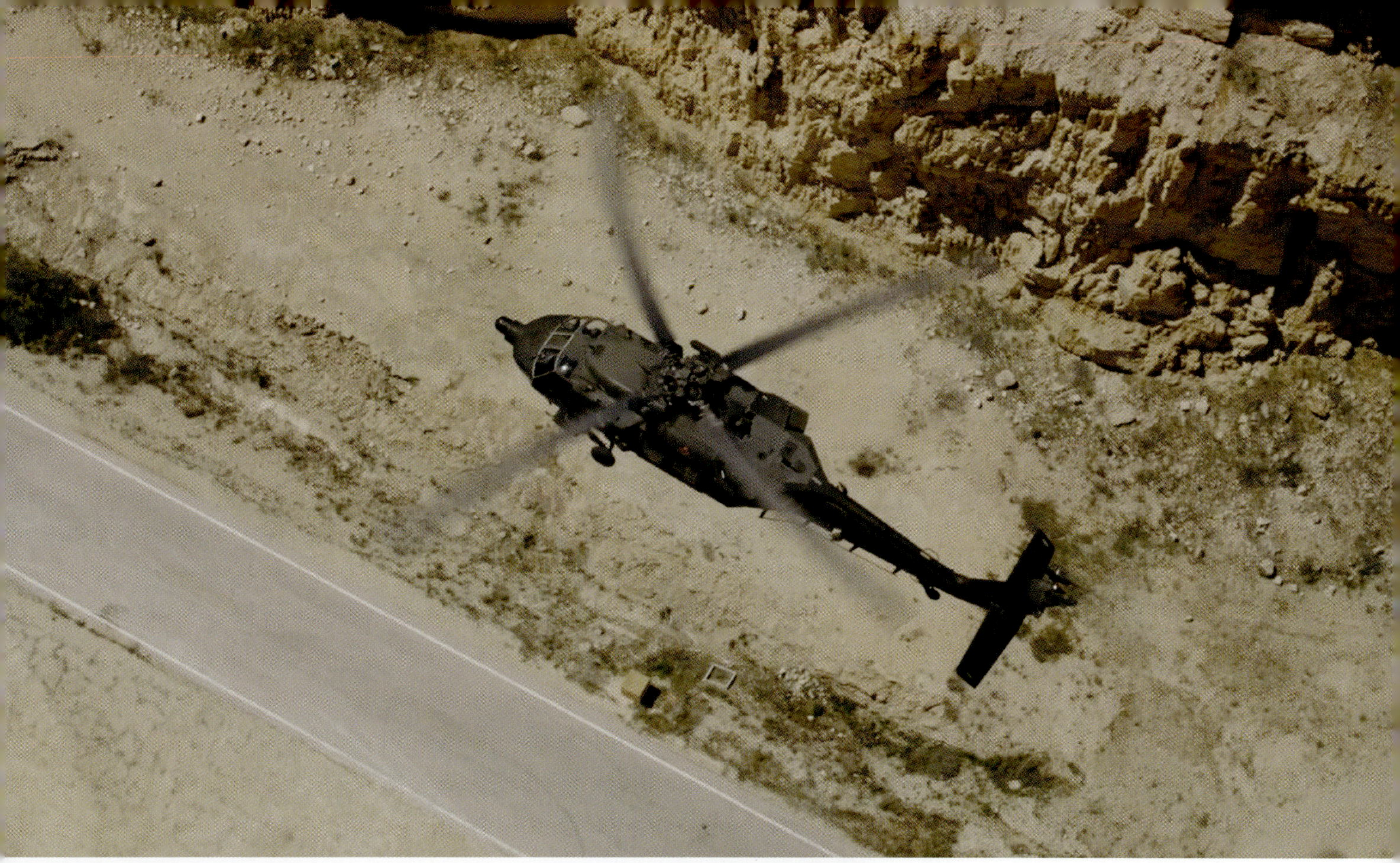

▲ Using a shutter speed of 1/50 sec. or 1/60 sec. will give a smoother sense of motion, as in this shot of a Jordanian special forces Black Hawk helicopter. Filmed by Lec Park (see pages 154–155) this drone shot was carefully coordinated with the helicopter's pilot—normally drones should steer well clear of other aircraft!

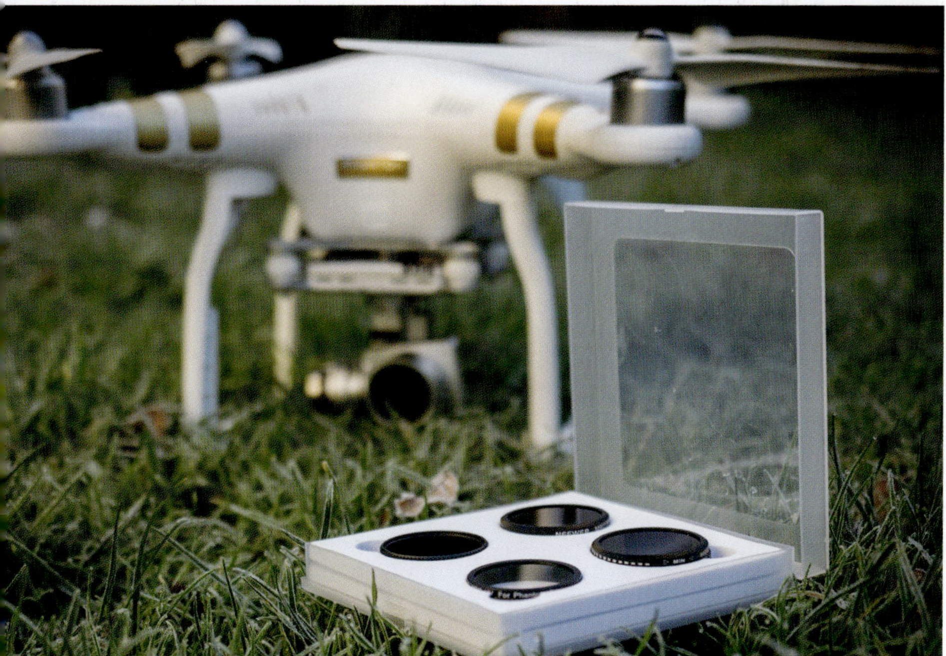

◀ A set of ND filters will enable you to set the desired shutter speed whatever the lighting conditions.

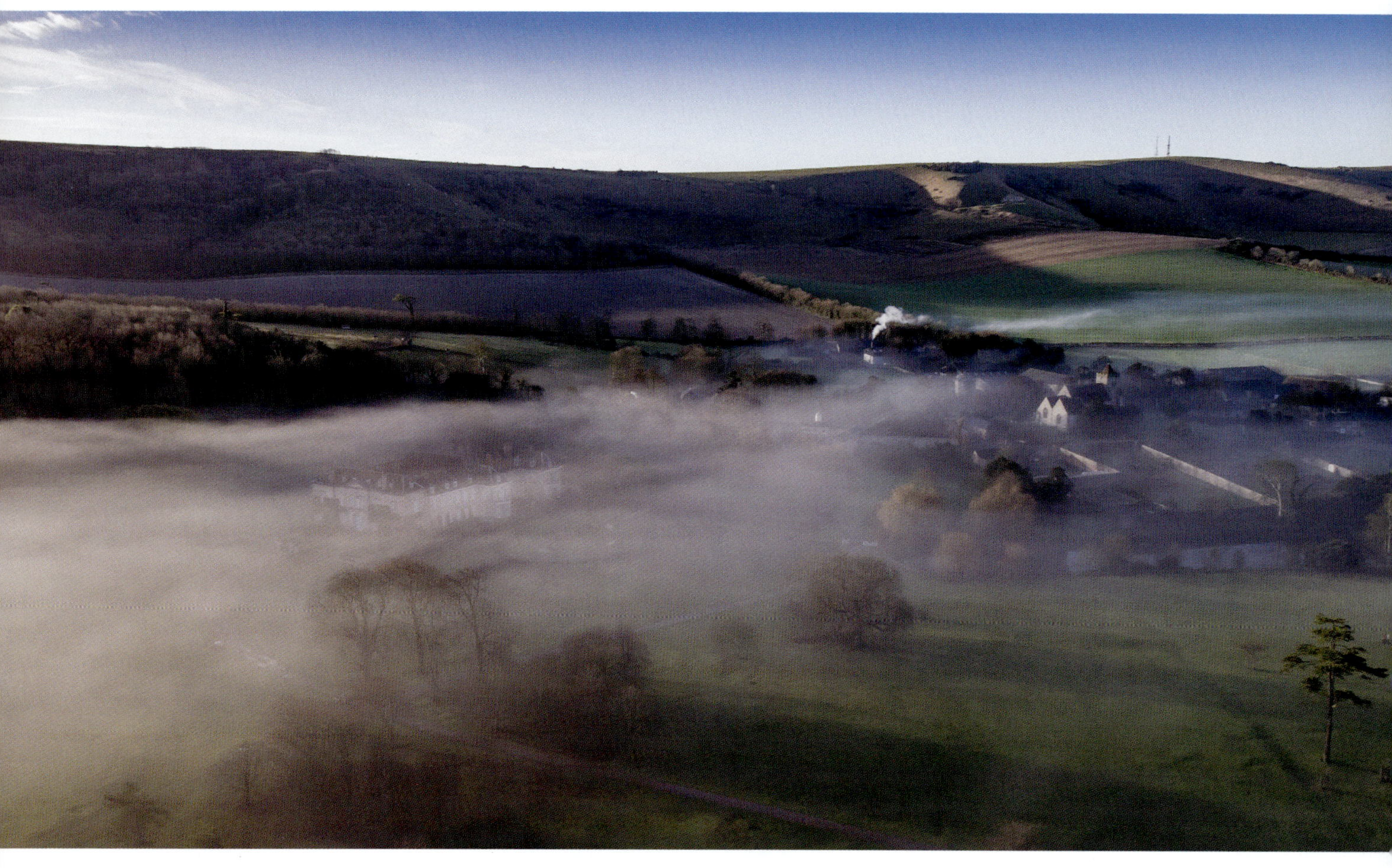

APERTURE

As discussed in the previous chapter (page 110), aperture refers to the size of the opening in the lens, with a wider aperture letting through more light than a smaller one. This is not the only effect of aperture, though: a wider aperture will also lead to a shallower depth of field (see page 111). Although this is not often used in aerial video, it is an effect that can be pleasing under some circumstances.

However, it's really only possible to achieve a noticeably shallow depth of field if you are using an interchangeable lens camera with a large sensor and a lens with a focal length of 50mm or longer. It also helps if the subject is close to the camera. Conversely, cameras with a small sensor and wide-angle lens (such as the GoPro or DJI Phantom cameras) will produce a deep depth of field.

ISO

ISO was explained on pages 112–113, and is the same regardless of whether you're shooting still image or video: in short, all cameras have a base ISO and if there is a good amount of available light, it is best to shoot at this setting for highest quality video. It is worth noting that ISO is sometimes referred to as "Gain" on dedicated video cameras.

▲ With most drone shots, the wide-angle lens and relatively distant subject matter will mean that the depth of field is extensive, so all of the image is in focus.

FOCUS

4K or UHD video is not uncommon in a drone, but with high-resolution video it's essential that you nail your focus. Thankfully, this is usually a fairly simple task and unless you are flying very close to your subject (at a distance of less than 5 yards/meters) you will usually find that you have sufficient depth of field to cover everything from that distance to infinity.

If you're using a drone with a fixed-focus lens you don't need to worry about focus at all, but some drone cameras—particularly those with larger sensors—do need to be focused. This is usually achieved by tapping the app to pick a focus point, or by prefocusing the lens manually before take off. If you're focusing manually with an interchangeable lens camera, it's a good idea to lock the focus once it has been set, to prevent any "focus hunting" from the camera's AF system. Better still, you can switch the camera and lens to manual focus mode instead.

SWEET SPOT

Assuming you have plenty of light and want a deep depth of field, it pays to choose your aperture carefully. Every lens will have an optimum aperture setting when it comes to delivering the sharpest images, which is typically in the region of f/5.6–f/8.

▼ In low light, you will have to raise the ISO substantially. Depending on the size and quality of your sensor, this may introduce noise or "graininess" to your video footage.

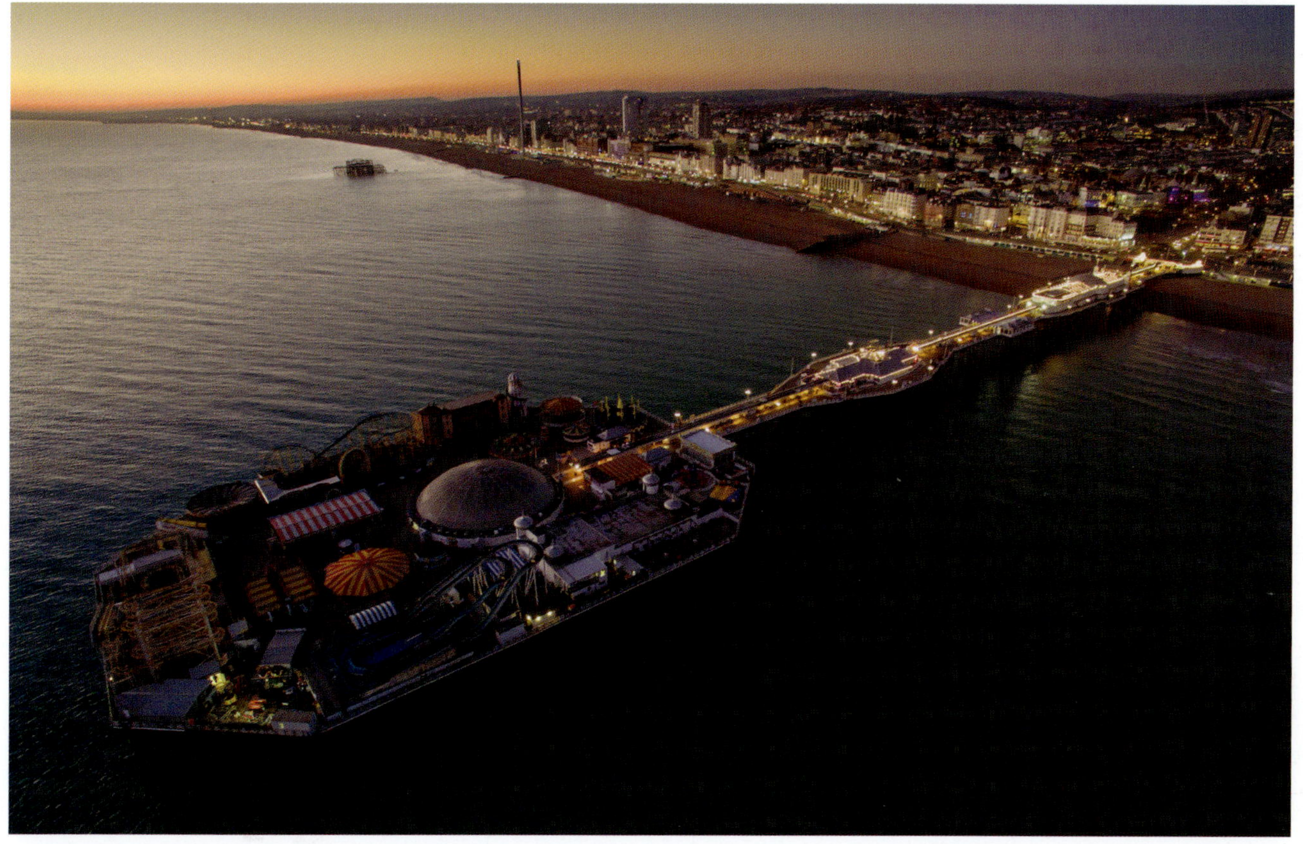

RESOLUTION, COMPRESSION & FRAME RATE

RESOLUTION

As with still photography, resolution refers to how much detail you can resolve in an image. One of the main factors here is how many pixels make up the video file. All other things being equal, it is usually best to go for the highest resolution setting available. For most modern consumer and professional drones this is "4K," or more accurately, Ultra High Definition (UHD), which produces video footage with a resolution of 3840 x 2160 pixels.

However, there is a number of reasons why you select a lower resolution instead. The first is compression. The image processing channel and write speed of your memory card will limit the rate at which data can be written. Many consumer drones have a maximum data rate of 60 Mbps, which requires pretty heavy compression for a UHD video. Consequently there may be times when you start to see blocky compression artifacts in the video, particularly when there is rapidly changing fine detail in an image (flying fast over a forest canopy, for example). In this situation, reducing the resolution to 2.7K—while maintaining the same data rate—gives a more pleasing result.

You may also find that compression artifacts that weren't immediately obvious start to become evident after color grading. So, if you plan on doing a lot of color grading, you might be better off shooting at a slightly lower resolution to preserve subtle color gradations and shadow detail.

Finally, if you want to shoot at a high frame rate (50 fps or above), you will likely be forced to reduce the resolution.

CODECS

When you shoot video on a consumer 4K camera drone, you're effectively capturing 8-million pixel resolution images, each with 256 possible colors (8 bit)—and this is usually happening 25–30 times per second. That produces an enormous amount of data to crunch!

The way a camera processes this data and compresses it is known as the "codec." The less compression there is, the better quality the image will be, and the more that can be done with it in post-processing. However, that comes at the expense of very large file sizes, which can quickly fill the camera's memory.

Codecs can work both spatially and temporally. In other words, when reading pixels along a row, the codec might say "the next 10 pixels are about the same," rather than storing the data for each individual pixel. Alternatively, the codec might say "this pixel is about the same for the next 10 sequential frames."

Depending on the way it works, the image from one particular codec can look very good if there is not much movement in the shot and there are large uniform areas in the frame, but might struggle when your drone is moving over an area with lots of details. This will be manifested as compression artifacts, which are areas of "blockiness" or smudged detail. To avoid this, you should use the highest quality video settings available on your camera—it will fill up memory cards quicker, but it's worth it.

Some codecs are also more efficient than others and can store good-quality data in relatively small file sizes. However, they might take a lot of processing power to decode at the edit stage, making the video play jerkily while you are editing it.

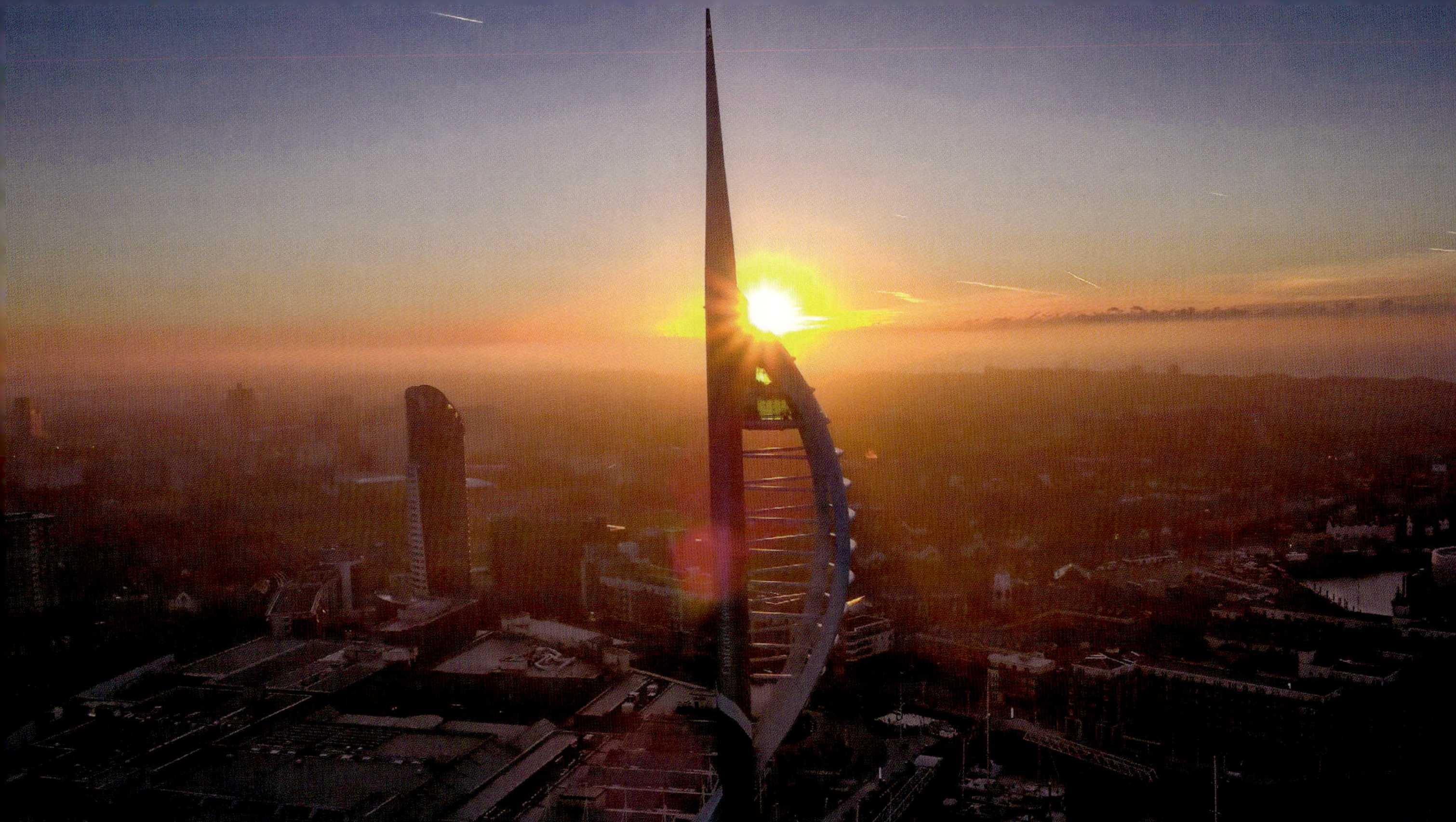

▲ Sunrise over Portsmouth Harbor, UK. Better codecs will give smoother color gradations in the sky, but this comes at the expense of more storage space on media.

FRAME RATE

Video frame rate refers to the number of still frames recorded per second. Usually, the frame rate you shoot at will match the frame rate you intend to play the footage back at, which will typically be 30 fps in parts of the world that use the NTSC TV standard (notably the USA and most of Latin America and Japan) and 25 fps for the PAL standard (Europe, Africa, China, Russia, and much of the rest of the world). Cinema movies are often shot at 24 fps.

However, you don't need to stick to these rates: you can also use faster and slower frame rates for creative effect.

Overcrank

"Overcranking" is when you run your camera at a faster-than-standard frame rate. For some applications you may want to shoot at a higher frame rate (such as 60 fps) and play the footage back at the same high rate. This gives smoother motion and is already compatible with YouTube (although this standard is yet to be adopted by the majority of TV broadcasters).

Overcranking can also be used if you want to achieve a slow-motion effect: if you film at 60 fps and play the footage back at 30 fps, for example, you will create "half speed" slow motion. Increasingly, relatively inexpensive cameras are able to shoot slow–motion video (the latest GoPro camera can shoot full HD video at up to 120 fps, allowing playback that's 4–5x slower than "real time" footage).

Perhaps the most obvious use of shooting higher frame rates is to record fast-moving action and sports. If the action is shot at 120 fps, for example, most of it can be played back in real time, but then slowed for particularly spectacular moments. However, due to the limits of data processing it is worth noting that it is usually not possible to shoot at a high frame rate *and* high resolution, unless you have a top-end camera.

Undercrank

Just as you can increase the frame rate you shoot at, you can also shoot at a lower frame rate. When the footage is played back at a "normal" speed it will appear to have been sped up, creating a classic timelapse effect.

The precise rate at which you shoot will depend on your subject. If your subject is very slow moving—clouds or fog forming and dissipating, for example—you might want to shoot just 1 frame per second and then play the footage back at 25 fps to speed it up by 25 times.

Timelapse effects can also be used to record over a period of months or even years. For example, recording a few seconds of video or a single photograph of a construction site every day or week over the course of the building work could create a dramatic timelapse movie. Of course, this would rely on getting the drone in exactly the same place each time, but there are several third-party apps that can record not only the position and altitude of your drone, but also the orientation of the camera, and repeat it precisely (within the limits of GPS accuracy) and at the press of a button.

▶ Fog or clouds make great subject matter for timelapse videos, just remember to keep any movement of the drone very slow and steady.

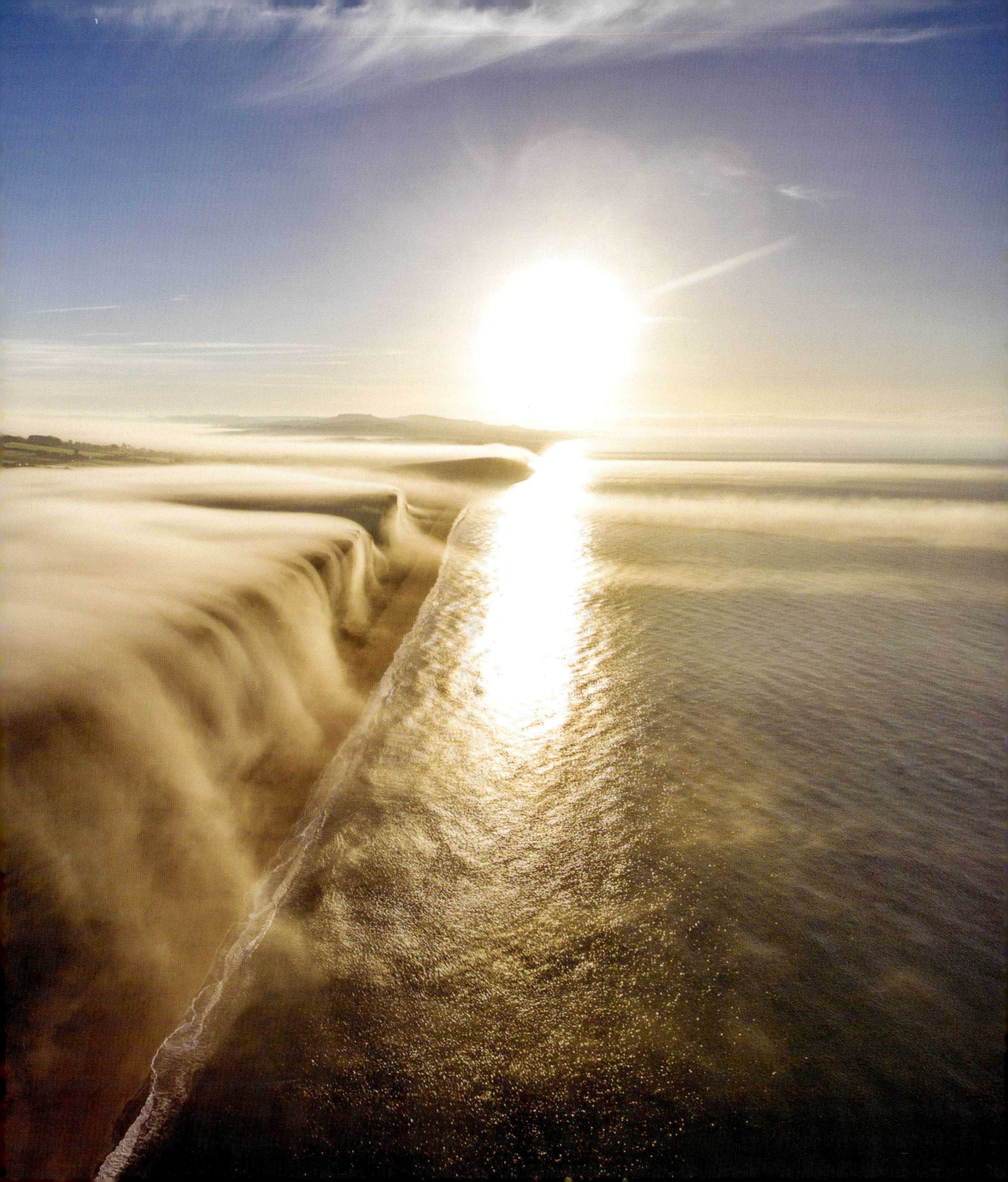

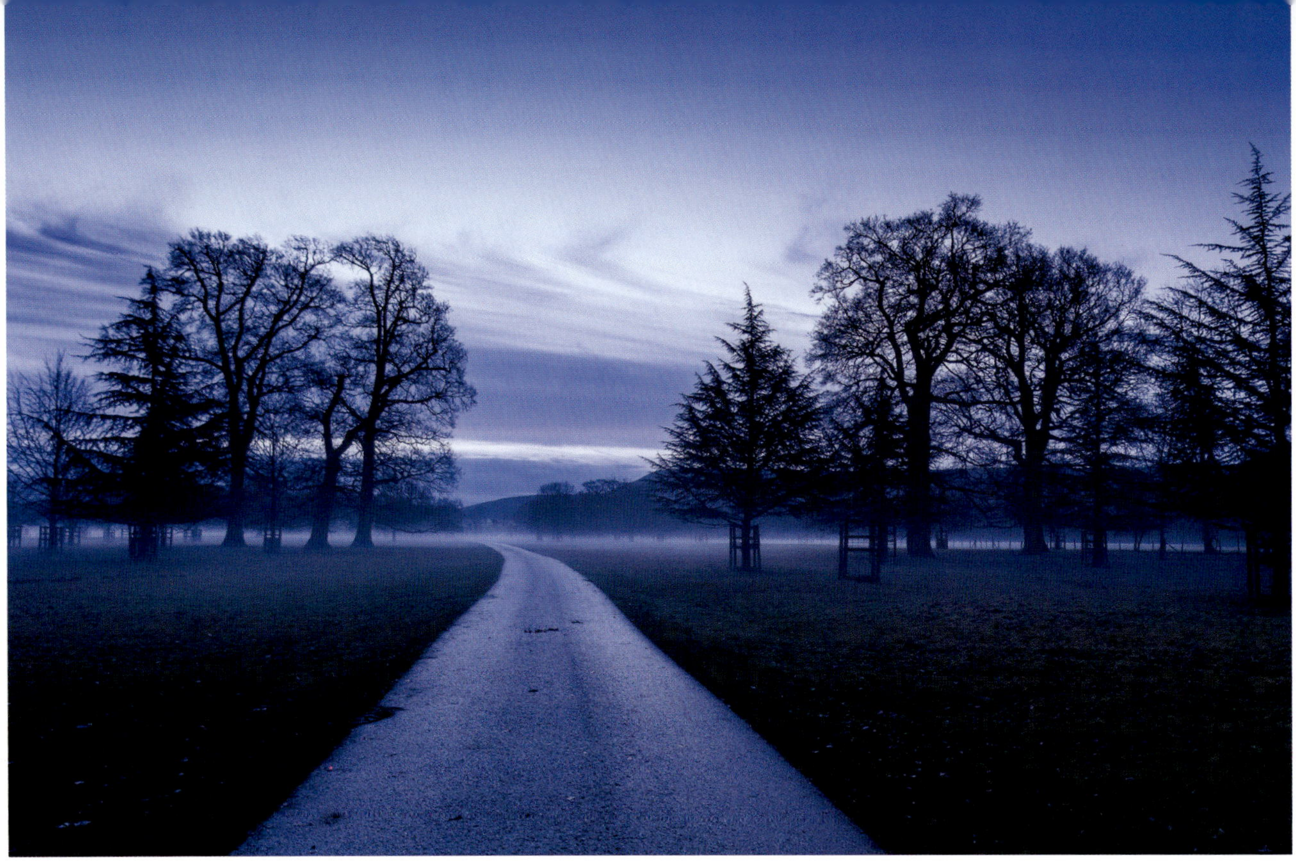

COLOR

WHITE BALANCE

White balance is the process of correcting the colors in an image to account for the color temperature of the ambient light. Most of the time when you are flying outdoors in daylight, your camera's Auto white balance setting is all you will need. However, if the colors in a scene change, the camera can make a noticeable color adjustment, which can affect your footage.

For this reason it is usually better to set the white balance to a Sunny or Overcast setting (most cameras offer a range of preset white balance options to choose from). If you are filming indoors under artificial light you might need to select Tungsten (Incandescent) or Fluorescent for best results.

You can also use your camera's white balance for creative effect: deliberately setting the "wrong" white balance can introduce strong color casts that give images a distinctly "warm" or "cool" look.

▲▼ The white balance is usually set to match the color temperature of the ambient light (below), but it can also be used to achieve a particular look. Here, raising the color temperature gives a cool blue look to the scene, emphasizing the idea of a physically "cold" temperature (above).

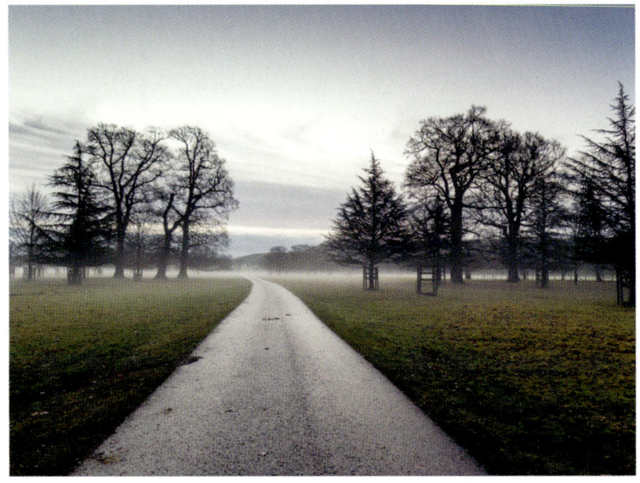

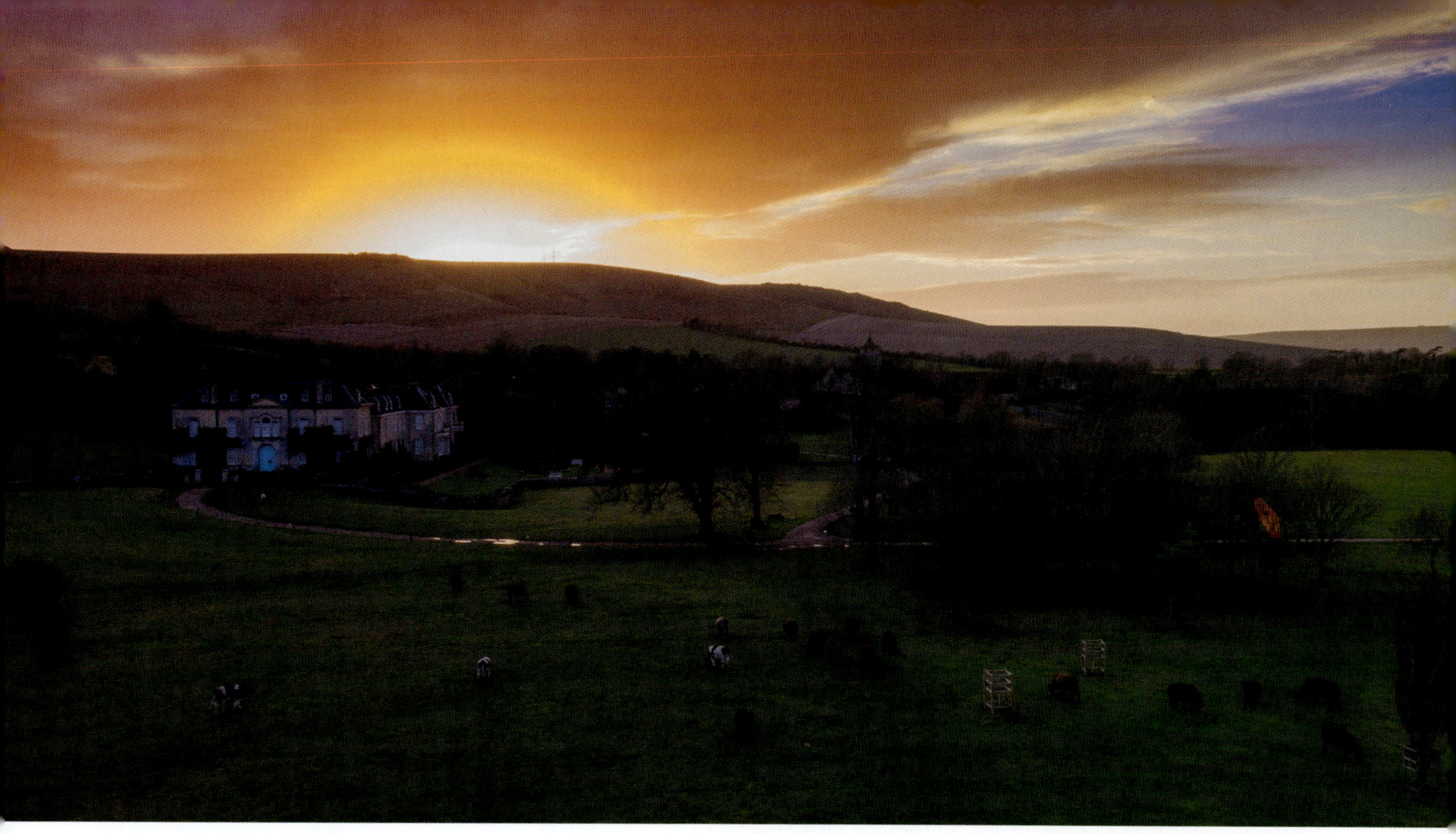

PICTURE STYLES

For the best possible results you should aim to "grade" your footage on your computer after you have shot it, as this gives you the most control over the final look of your footage. Assuming you are going to grade, the first thing you need to do is to shoot in a format that preserves the maximum amount of digital information and therefore gives you the most flexibility in post-production.

If you plan on making large adjustments, Raw video is the colorist's dream. Essentially you are recording all the data from the camera sensor, so each frame is like a Raw photograph (see page 118), with the option to adjust colors, bring up the details in shadows, or put your own look or style into your footage. However, the option of shooting Raw video is currently only available on some higher-end cameras and drones.

On most other cameras, you have to rely on a carefully chosen "picture style" for your video footage, which sets the level of contrast, color saturation, and sharpness in-camera. Many camera operators recommend shooting with a "flat" picture style that delivers low contrast, low color saturation, and low sharpness, as it's easier to increase the values of those parameters during post-production than it is to decrease them. But with highly compressed footage—such as the 4K resolution at 60 Mbps that you find on many consumer drones—I would hesitate to go too flat, as the compression may not allow a very large contrast increase. Often, a moderate profile is a more appropriate starting point.

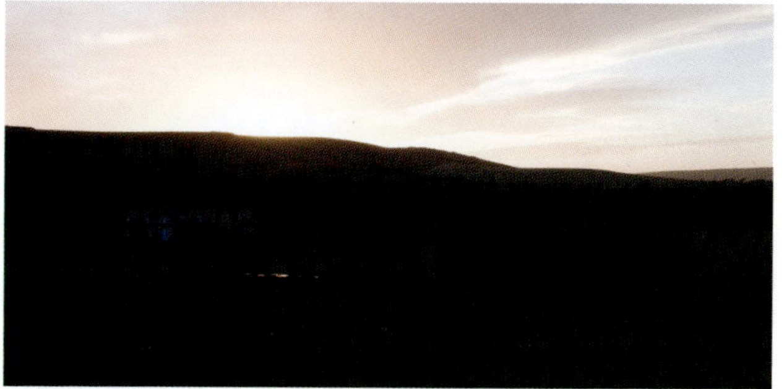

▲ Color grading can have a huge effect on the look of your video footage. For the greatest flexibility it's best to shoot Raw or at a very high bitrate.

IN THE EDITING SUITE

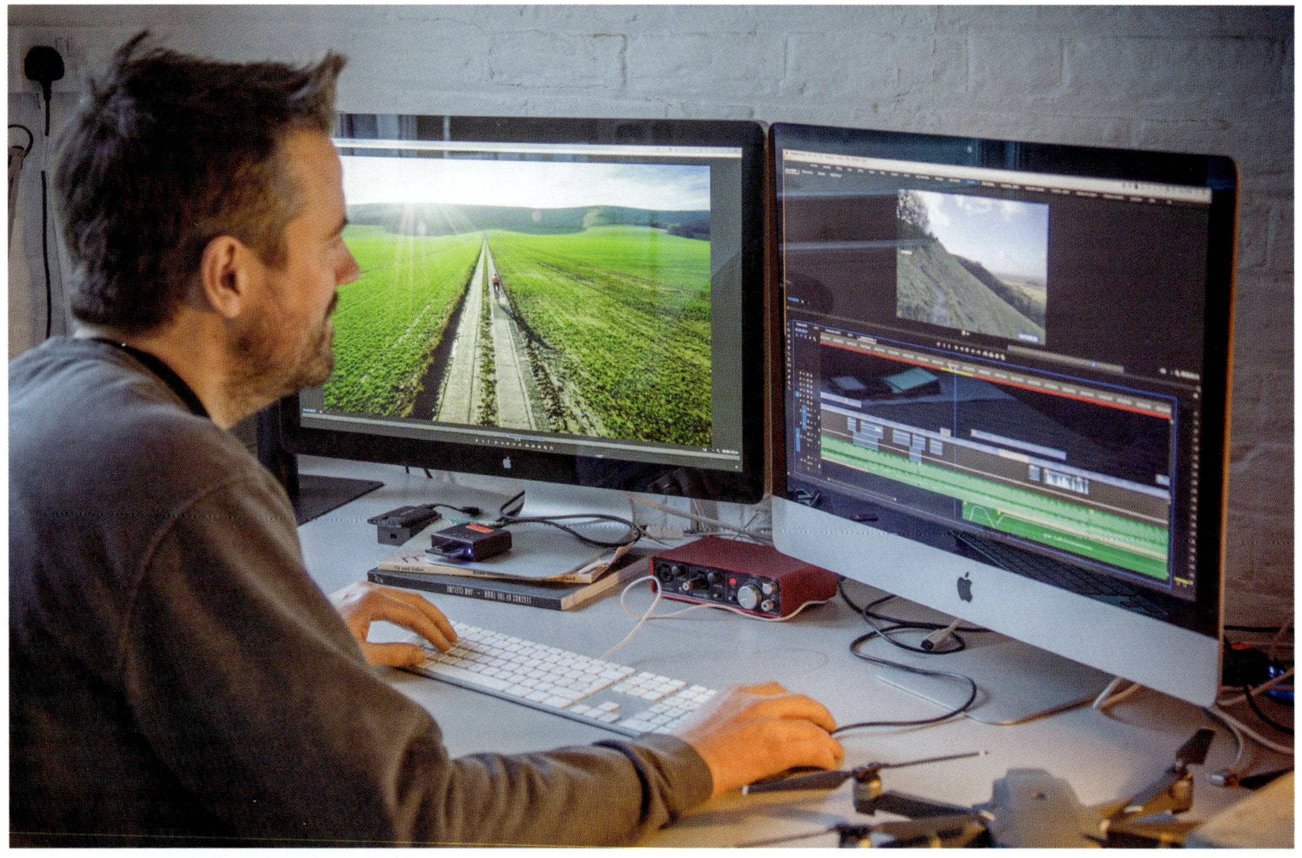

▲ **Becoming skilled at editing and color grading will not only enable you to show your footage in the best light, but it will also influence how you shoot when you're out flying.**

This book is not supposed to be a detailed "how to" on video editing: there are plenty of excellent books dedicated solely to that topic. Instead, in this section I aim to share a few tips and ways of thinking that I have found helpful over the years.

It often feels like a real treat to open up some great aerial footage you've just shot and see it in all its glory. If you are working commercially on a large shoot, it is unlikely you will be the person editing and color grading your shots. In this case, your responsibility ends with handing over your rushes (unedited clips straight from the camera media), which will hopefully be perfectly exposed, in focus, and with the correct white balance, frame rate, resolution, and codec.

Ideally, you will have planned each of your shots and where they go in your story, before you started shooting. This could be a second-by-second script with a detailed description of each shot, or it could be a pictorial storyboard. In either instance it will give you a blueprint from which to work during the editing process. Inevitably, you will deviate to some degree from the original plan as you go along, but it really helps to have some guidelines. This applies whether your film is a music-driven short or part of a full-length narrative film.

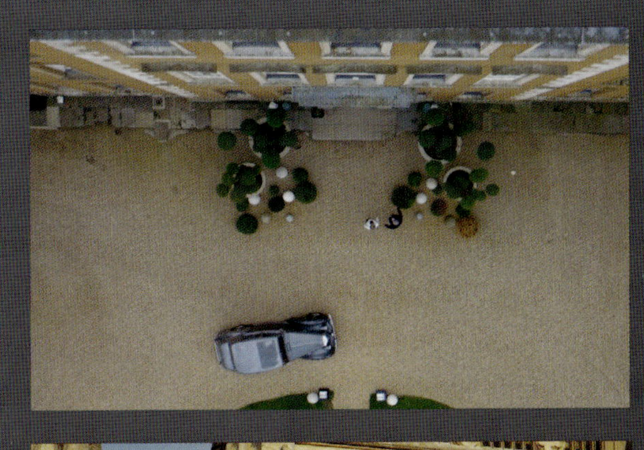

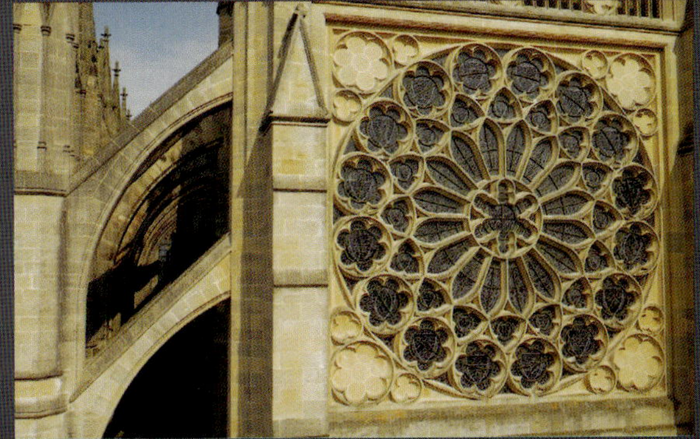

YOUR SHOWREEL

You might want to edit some highlights of your best drone footage to put up on your website or social media. This is a fun way of showing off your skills as a pilot and camera operator. The following tips will help you create a strong video showcase:

- Choose music to fit the mood.
- Keep your showreel short (under 3 minutes is ideal).
- Put your best material at the start—viewers will often stop watching after the first 10 seconds.
- Keep the pace snappy.
- If you're a commercial operator make sure you include your contact details.
- Keep a copy of your showreel on your phone so you can show potential clients.

INITIAL STEPS

If you are editing your own footage, the first thing you will need to do is to transfer the files from the camera media to a hard drive (you should also make an additional backup copy, just in case you suffer a hard drive failure).

Next, you will probably open a new project in your editing software of choice and import your rushes. How you organize your files within the project will depend on personal preference and your software, but it's a good idea to use a system that makes it easy to locate any given clip in the future. This is particularly true if it's a large project with a lot of footage.

Scrub through the rushes and identify the parts you need. Mark them with "in" and "out" points, possibly dragging each section onto the timeline to create a rough edit. As you work, think about how one shot leads into another, and how each shot helps with the story you want to tell. Ask yourself if the cuts work. If a cut between shots jars, look at the timing, the motion, and the composition of the footage either side.

The art of making a "good" cut takes practice and experience, and a good editor will have a keen instinct for what works. Often, removing (or adding) a single frame at the start or end of a clip can make the difference between a successful cut and one that doesn't work.

If your film is music-driven, choose your track at the beginning of the process and lay it on the timeline at the start. Audio has a huge influence on whether a cut is working or not, so make the most of the music, changing your pace to work with the soundtrack: cut to the beat when you want to emphasize the rhythm of the song, and place key cuts on key parts of the song to make the most of any emotion or change of pace.

Be wary of cutting right on the beat too much, though, as it runs the risk of every shot being exactly the same length, which will quickly become monotonous. The length of each shot should be tailored to not only its content, but to the ebb and flow of the film and the message you wish to communicate. If you want your film to feel stressful, frantic, and energetic, create a staccato effect by cutting short clips together; for a more serene or majestic feeling let shots play longer.

PROFILE: LEC PARK

Lec Park is a highly experienced drone operator working primarily in the TV and film industry. As a former professional mountain biker, Lec got the bug for film making via extreme sports filming. Having been on the cutting edge with small action cameras and cable-cams, taking to the skies with a drone was an obvious next step. His work has since taken him all over the world, working on a wide variety of high-profile TV, and commercial video projects.

Having worked with drones since the very earliest days, Lec has seen how the technology has developed. It used to take a lot of skill and experience to build and correctly set up a camera drone, making it something of a specialist occupation, but today anyone can achieve technically good, stable footage from a modern ready-to-fly drone. This has meant that the emphasis has shifted away from the purely technical side of flying a drone, and successful commercial shooting now requires more imagination and an eye for a great shot.

▶ For the opening shot for the BBC's car show, *Top Gear*, Lec used a DJI S1000 drone fitted with a Panasonic GH4 Micro Four Thirds camera. At the time of the shoot, the weather conditions were terrible, with intermittent rain, and very strong winds. As there was a headwind, the maximum speed the drone could manage was 20mph (roughly 30kmph), so this was the speed the cars had to maintain. Despite this, the greatest challenge came from trying to get the cars to hold their formation for the duration of the shot.

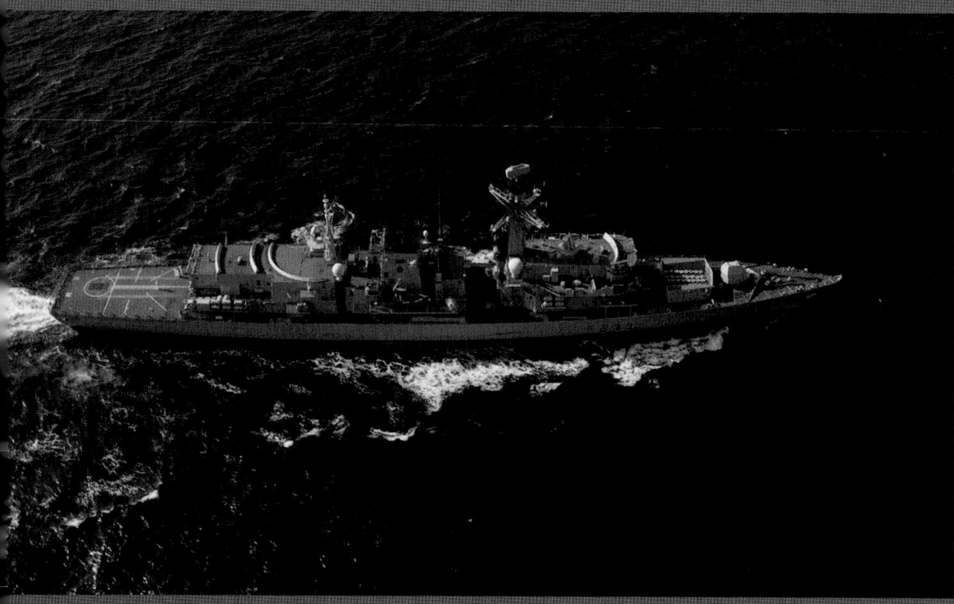

◀ To film a live-firing exercise as part of Amazon's *Grand Tour* series, Lec used a DJI Inspire Raw with an Olympus 12mm f/2 lens. Not only did he have to contend with strong headwinds and crosswinds, but he also had to take off from and land on the ship,HMS Richmond—which was traveling at 15 knots—as well as avoiding an exclusion zone around the muzzle of the gun.

▶ Lec used a DJI S1000 drone with a Panasonic GH4 camera and 12mm f/2 Olympus lens to shoot a stunt in which two aerobatic planes flew through a hangar. This was a tremendously complicated shoot, with a chase airplane, full-sized helicopter, and Lec's drone in the air simultaneously (as well as eight ground-based cameras in the hangar). The shoot was therefore planned in meticulous detail with regards to safety, and an observer from the Aviation Authority was present throughout.

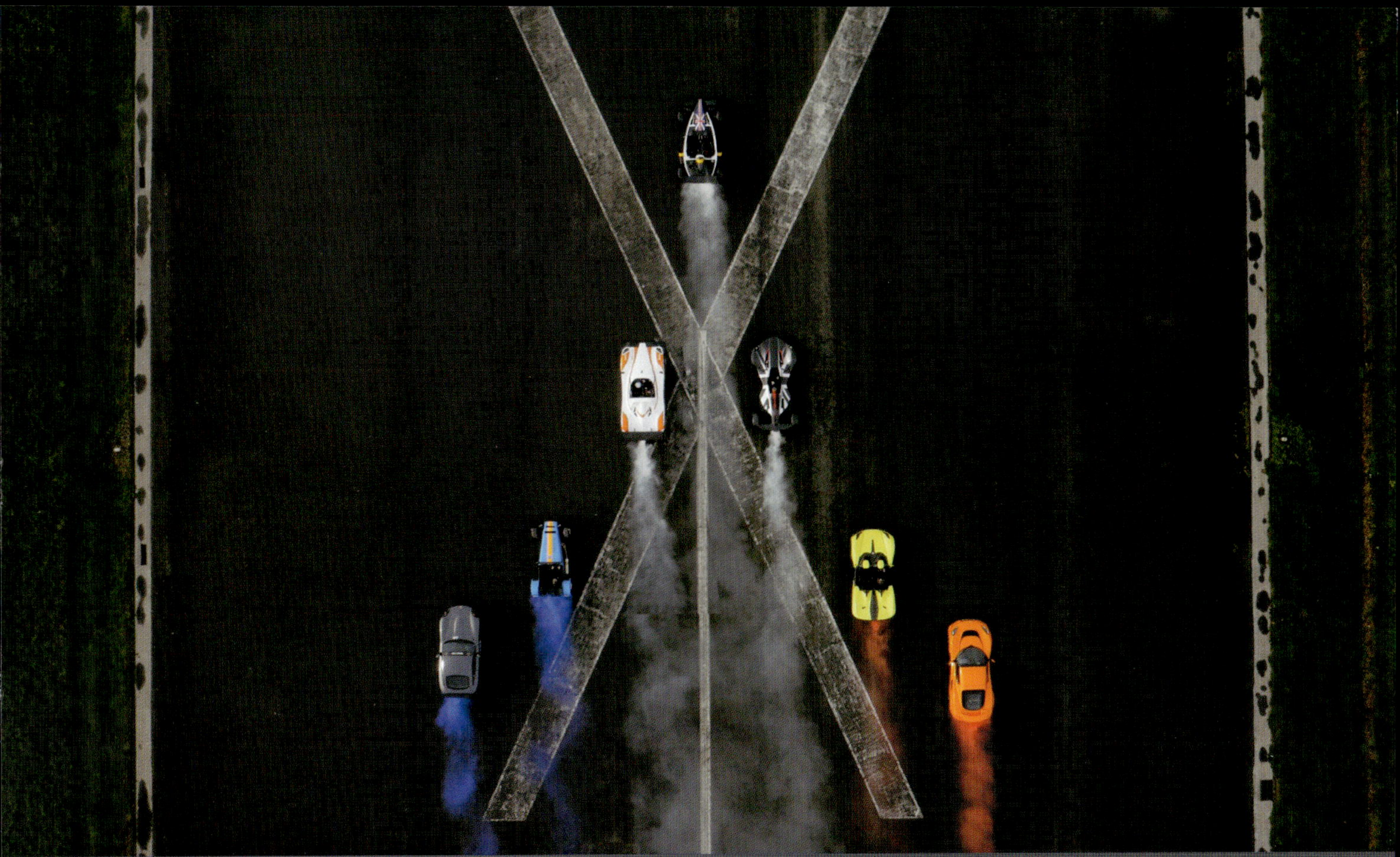

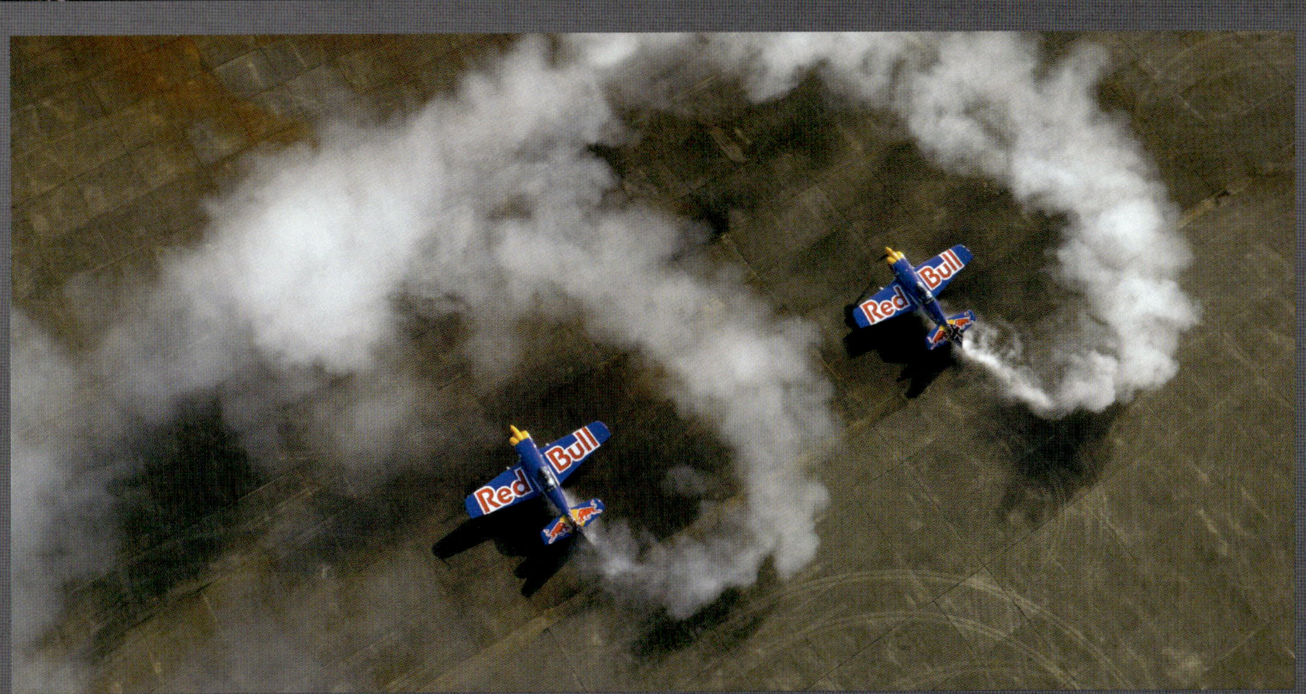
Red Bull
Red Bull

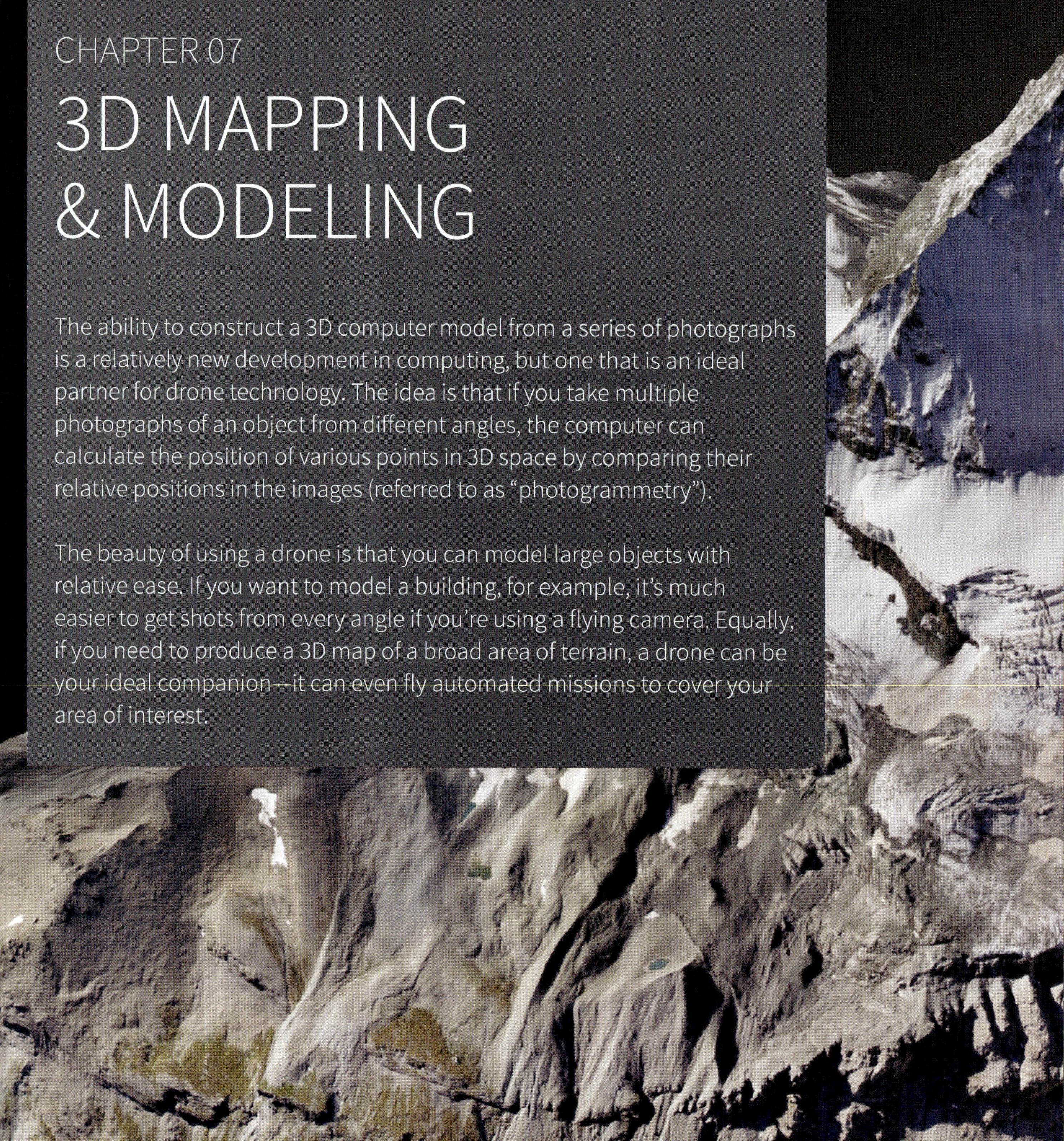

CHAPTER 07

3D MAPPING & MODELING

The ability to construct a 3D computer model from a series of photographs is a relatively new development in computing, but one that is an ideal partner for drone technology. The idea is that if you take multiple photographs of an object from different angles, the computer can calculate the position of various points in 3D space by comparing their relative positions in the images (referred to as "photogrammetry").

The beauty of using a drone is that you can model large objects with relative ease. If you want to model a building, for example, it's much easier to get shots from every angle if you're using a flying camera. Equally, if you need to produce a 3D map of a broad area of terrain, a drone can be your ideal companion—it can even fly automated missions to cover your area of interest.

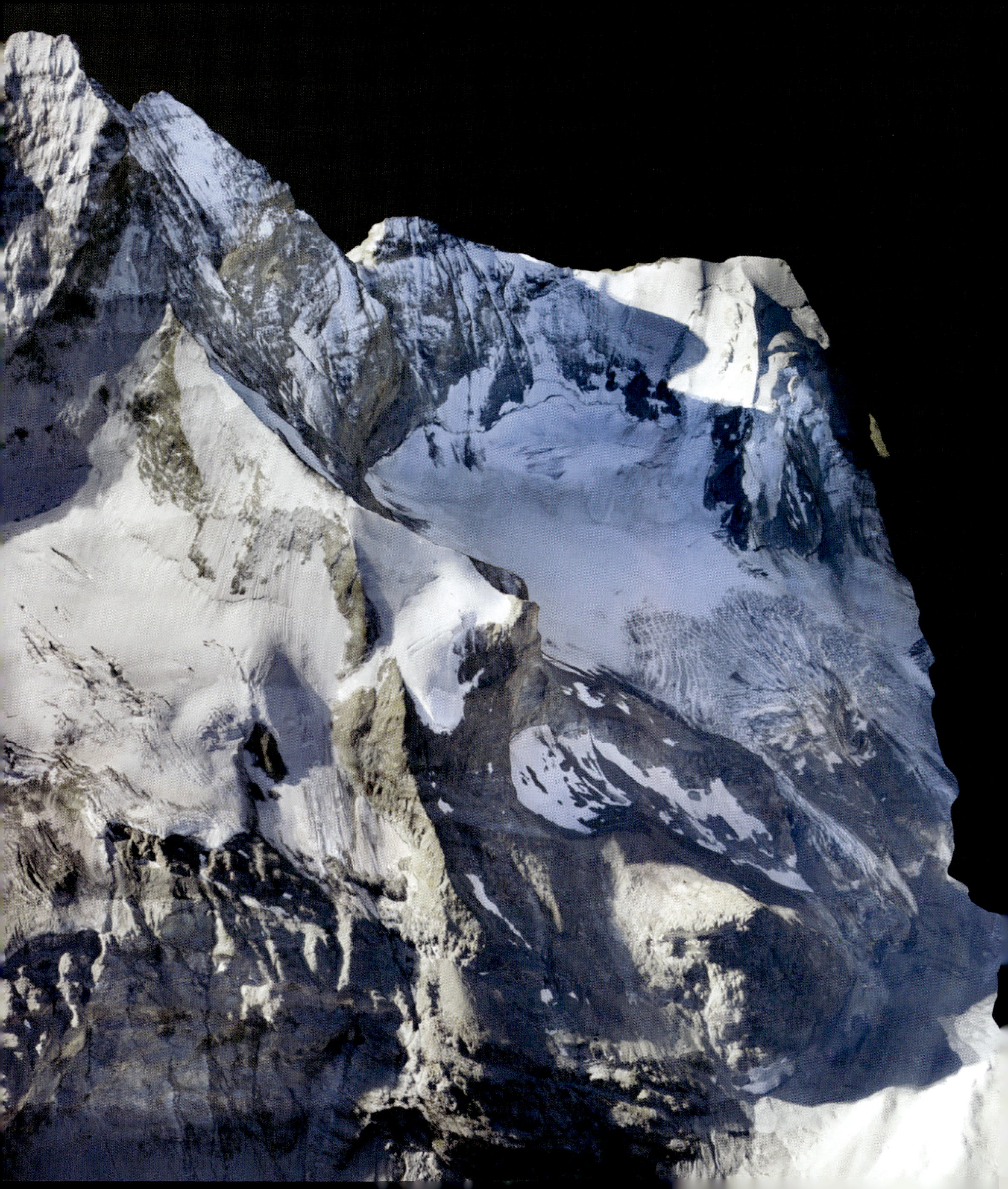

GETTING STARTED

WHAT DO YOU NEED?

Three-dimensional models and maps can be made with almost any camera and almost any drone: all you need is to take a lot of photographs from every conceivable angle and assemble them using dedicated computer software. There is a range of both free and commercial software for making 3D models from source photographs (see box). Some of the software is only compatible with certain popular drone models, but there is also software available that can ingest photographs from any camera.

GEOTAGGING

Many current GPS-connected camera drones automatically tag each photograph with its GPS location (latitude, longitude, and altitude). This information is particularly useful for mapping purposes, as your software can automatically "georeference" your 3D map—that is to say, each pixel on your map will have an associated set of map coordinates.

If your drone does not have the ability to geotag photographs, you can georeference a stitched map by identifying four or five "ground control points" that are easily identifiable on your aerial map and finding their coordinates. This can be done either on the ground with a handheld GPS unit (or the GPS on your smartphone), or by getting the coordinates from Google Earth.

In the past I have even gone to the lengths of attaching a small GPS-logging device to my drone (after syncing the time code with the camera) and using a third-party app to geotag the resulting photographs after the event.

SOFTWARE

123D Catch: Generic 3D modeling software. Free for Windows PC, Android, Windows Phone, and iOS devices.

Agisoft: Professional 3D mapping and modeling software for Windows or Mac. Various licenses available.

DroneDeploy: An end-to-end, cloud-based solution that includes an app to plan and fly your mapping mission. Works on Windows or Mac, with an app for Android/iOS.

DroneMapper: 3D mapping software for Windows that works with DJI Phantom or Inspire range. Has both free and paid-for versions.

Pix4D: A range of commercial mapping and modeling solutions tailored to support specific drone models. Includes mission planning and flight control app. Windows options at a range of prices; Mac support at Beta stage.

Skycatch: Includes an app to plan and fly your mapping mission. Processing is cloud based, so it works on both Windows and Mac, with an app for iOS. Monthly or annual subscription options.

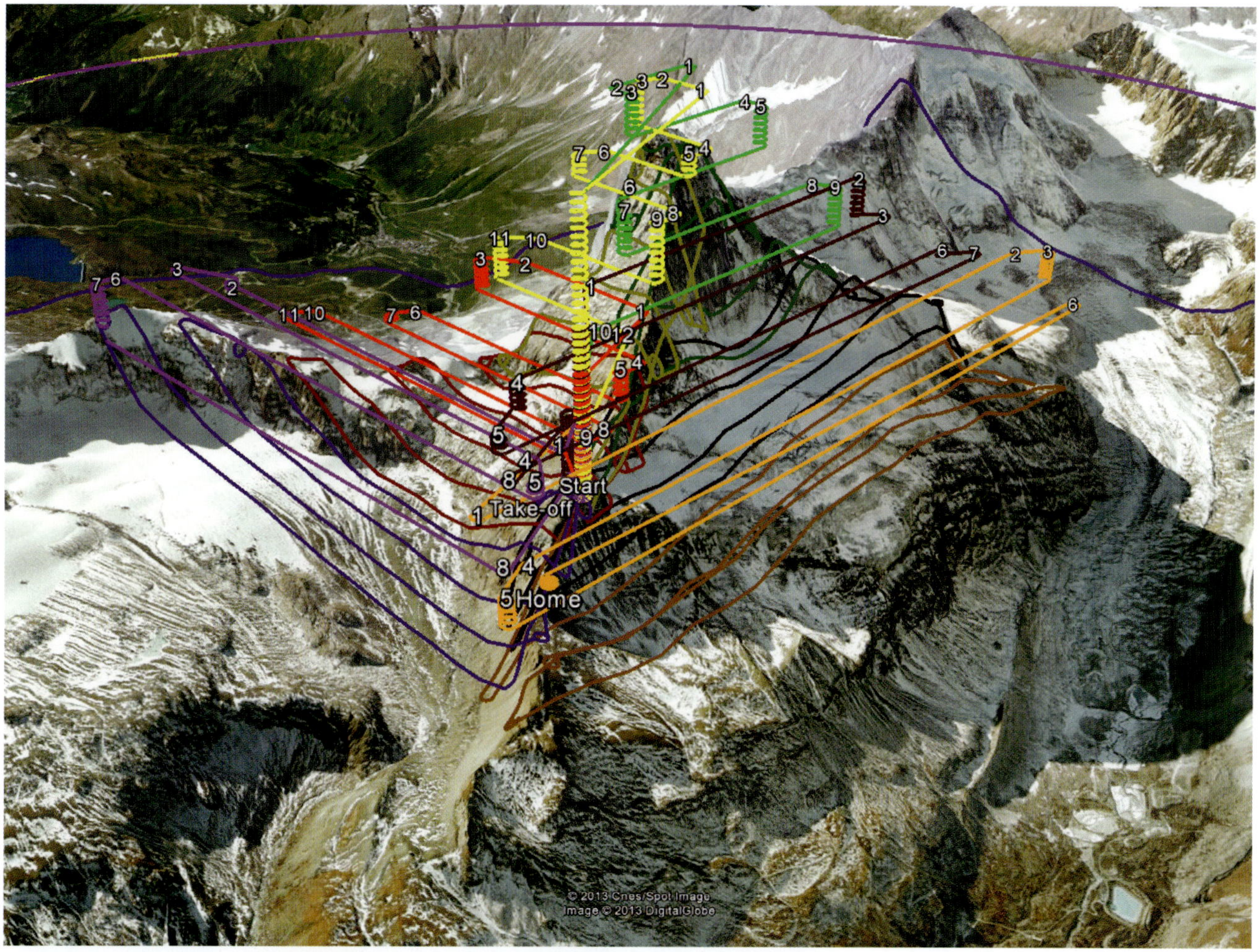

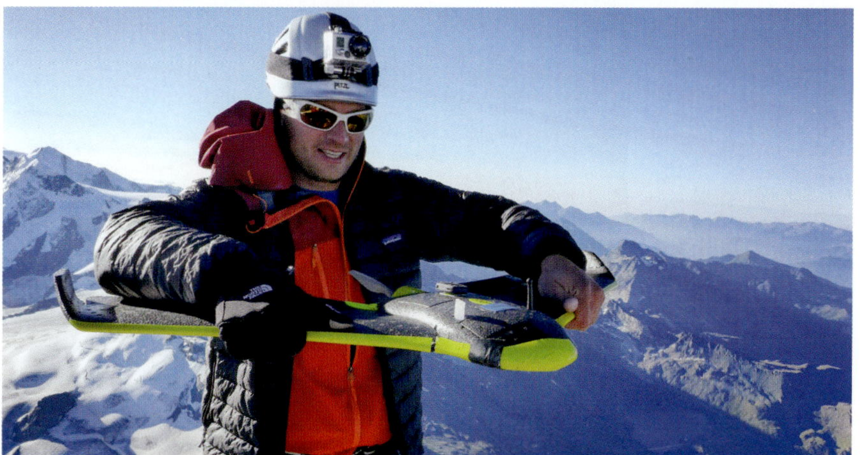

▲ When Pix 4D, Sensefly, Drone Adventures, and MapBox collaborated to make a high-resolution 3D model of the Matterhorn in the Swiss Alps, it was a hugely ambitious project. Meticulous planning resulted in a successful outcome, but it required 11 flights covering more than 164 miles (260km) and more than 2000 individual photographs.

◀ Launching a Sensefly eBee from the summit of the Matterhorn. These fixed-wing mapping drones coped remarkably well with the fickle and turbulent winds surrounding this magnificent mountain.

PLANNING YOUR MISSION

The ideal flight plan for shooting the source images for your model or map will depend on the shape of whatever it is you're mapping. For a largely flat area without significant vertical surfaces, a simple grid pattern with the camera pointing directly down should suffice, whereas steep or vertical surfaces may need some oblique shots as well.

▼ This 3D model of the Matterhorn in Switzerland was constructed as part of a cooperative venture between Sensefly, Pix4D, Drone Adventures, and MapBox. To produce it, 2188 images were taken over 11 flights with a fixed-wing Sensefly eBee drone.

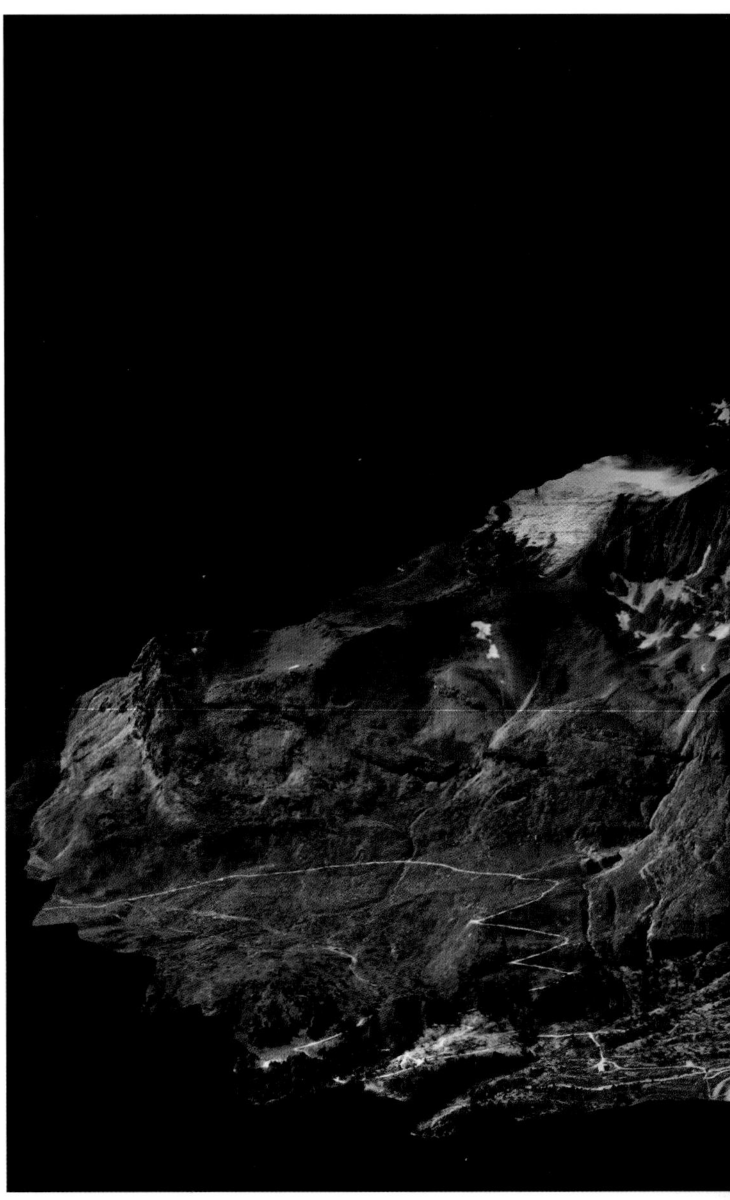

MAPPING THE FLAT

If the area you are mapping is predominantly flat, or undulating gently, you can shoot your photographs looking straight down. You can plan the mission with an app on your phone or tablet before you set out, which usually involves drawing a rectangle (your target area) on a Google Earth-style map. The app will then plan the flight path and determine how many photos it will take. It should also warn you at this point if the area is too large for a single flight (based on battery life).

Once you get to your location, all you should have to do is follow the app's instructions, upload the flight plan, and press "go." You will have various options regarding the altitude to fly the mission at and how much of an overlap you want between the shots. The drone should follow the pre-determined flight path—taking photos automatically as it flies—and then return to home and land.

If your drone model is not supported by any of the mapping apps, you will need to fly the mission manually, but the same basic principles apply. You are aiming to fly a grid pattern over your location, taking photos that overlap by around 60% in track (flight direction) and 50% in cross-track (between rows). In complex environments, such as areas with lots of trees or high buildings, use more overlap.

Fly your drone along an imaginary straight line with the camera aimed directly downward. Using your tablet or monitor view as a guide, move just under half a frame along the line, then take a new photo. When you get to the end of a row, move the drone across by about half the frame width, take another photograph, and then start to fly back, repeating the process with another row of shots that overlap by around 60%.

MAPPING TERRAIN WITH STEEP SECTIONS

If your terrain has steep areas, flying a grid as outlined previously may mean that the vertical or very steep sections of the terrain lack detail. This is where you can add to the source images with a "freeflight" mission. In addition to the orthogonal (downward-looking) images, shoot more detail on the steep or vertical areas by taking shots with the camera at an oblique angle, remembering to overlap your shots by 50% or more. These sections can be flown manually, either triggering the camera yourself, or using an app that will automatically fire the shutter at appropriate intervals, based on the distance traveled.

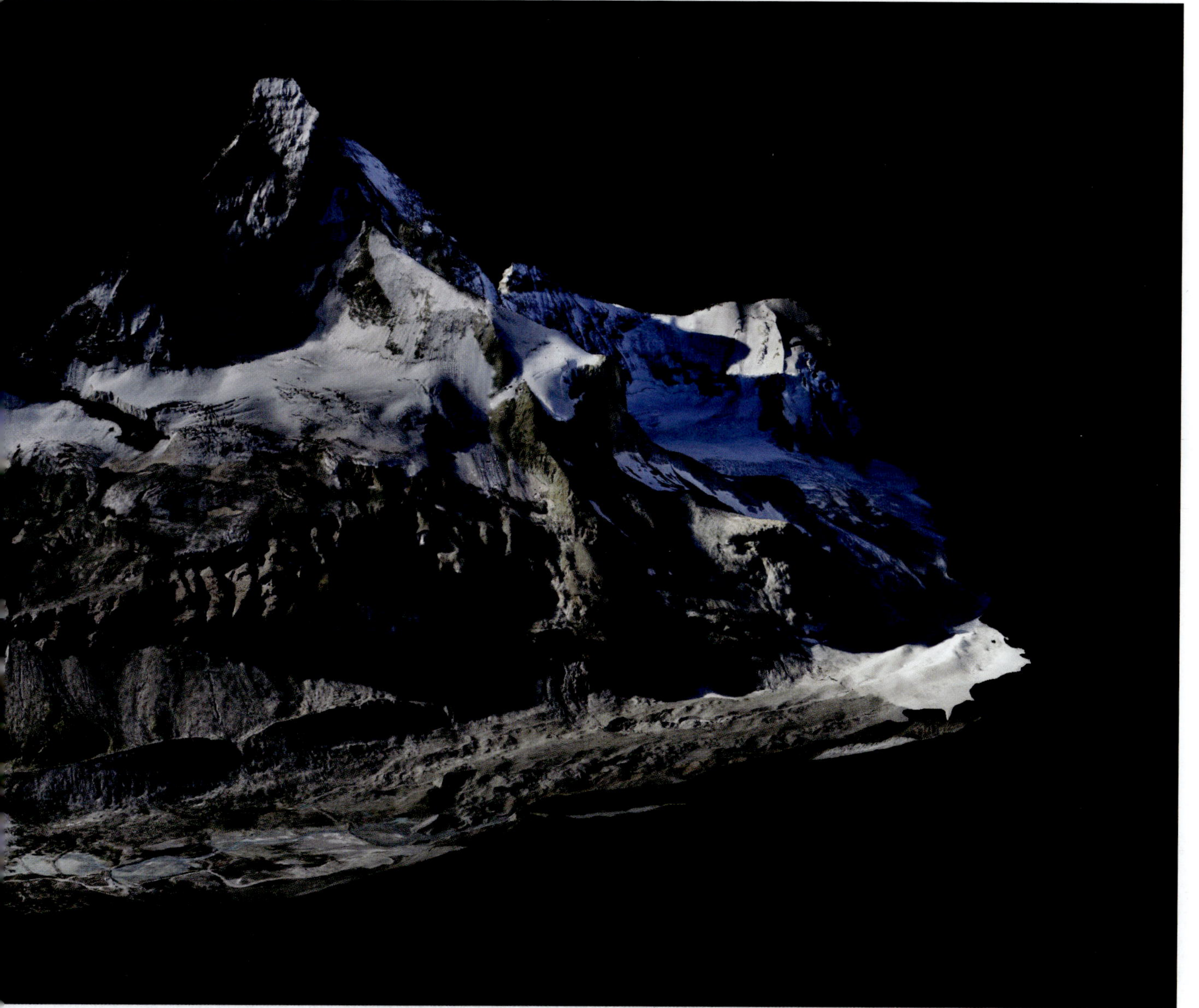

MODELING A BUILDING

If you want to make a 3D model of a building, it is usually best to fly manually and then either take photographs manually based on an overlap of 50% or more, or use an app to trigger the camera. A typical flight path might be similar to that shown in the diagram below.

Once you have completed your flight you will need to upload the photographs from the drone onto you computer to process the images. This processing will either be performed directly in the software or remotely in the cloud, and once it has been completed you are likely to have several output options to choose from:

Point cloud: A collection of points in 3D space representing positions on the surface of the mapped object. The point cloud can be converted into a "mesh" for 3D modeling purposes.

Elevation map: A 2D representation with heights represented by color classes or contours.

Orthomosaic: A georeferenced, ortho-corrected 2D map.

Digital Surface Model (DSM): A 3D representation of the earth's surface, including buildings, trees, and so on, constructed by draping an orthomosaic over a surface mesh.

Digital Terrain Model (DTM): Like a DSM (above), but shows the underlying terrain *without* buildings and trees. Collectively, DTMs and DSMs are also referred to as Digital Elevation Models (DEMs).

Once you have completed your 3D model or map, there are various ways that you can export and share it. Depending on your choice of software, you might keep it in the program's native format to share with friends or colleagues who use the same software. There are also options for exporting to technical formats for mapping, GIS (Geographic Information Systems), and CAD (Computer Aided Design) packages.

▲ A typical flight pattern for modeling a building might include two complete circuits at differing altitudes with plenty of overlap between images.

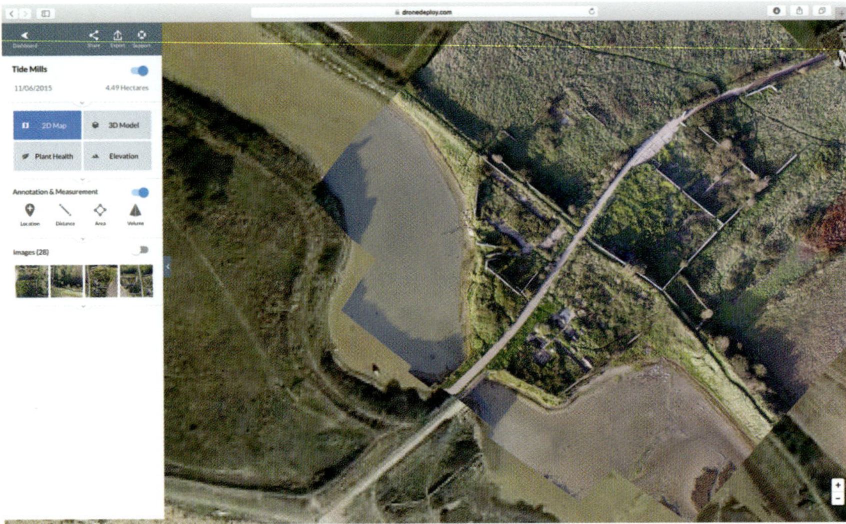

▶ An orthomosaic is a georeferenced 2D digital map. It is usually a composite of a number of individual photographs that have been orthocorrected to correspond to a particular map projection.

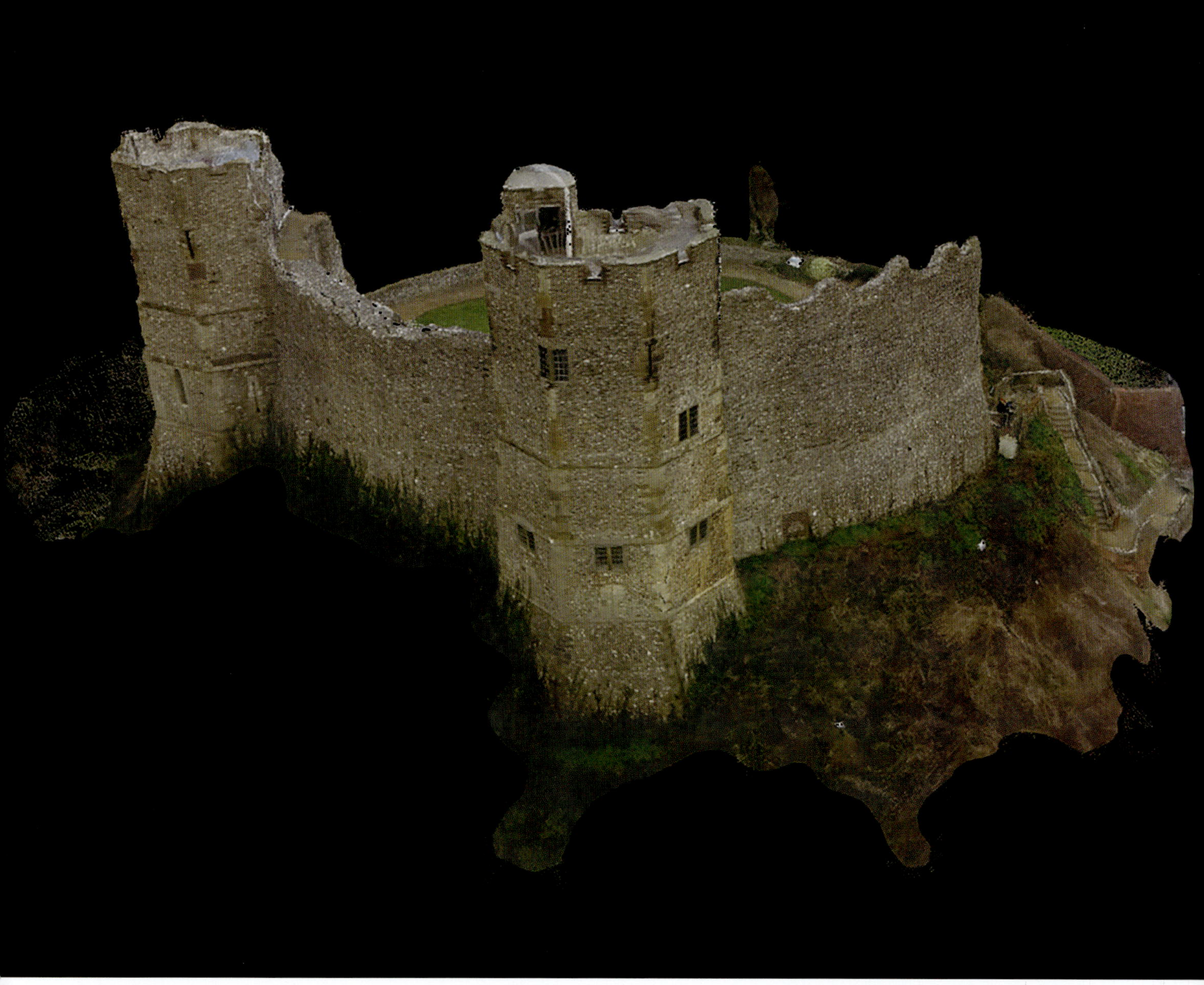

▲ This 3D model of Lewes Castle in Sussex, UK was created with the permission of the owners. The model was generated from 32 still images shot with a GoPro 3 camera on a DJI Phantom 2, which were then assembled using Pix4D software.

Alternatively, 3D models can be shared on dedicated websites for 3D models, such as www.sketchfab.com. Sketchfab models can then be embedded in your own website if necessary.

You can also record virtual flights through a 3D model as a video file. This can be done within some mapping/modeling programs (Pix4D, for example) or you may need to do a screen recording. The video file(s) can then be played using general video playing programs, or uploaded to video-sharing websites.

CASE STUDY:
MAPPING THE PICOS DE EUROPA

Researchers approached me recently from Qatar Carbonates and Carbon Storage Research Centre (Imperial College London). They had a project to make geological maps of an area in the Picos de Europa, Northern Spain, and had decided that using drones would be a useful supplement to more traditional mapping methods.

The planning phase involved careful examination of the area of interest on maps and Google Earth. We needed to make a realistic estimate of the surface area we could hope to cover during the fieldwork, based on the remoteness of the area and difficulty of access; the number of drone batteries available; the spatial resolution needed (and therefore height above take off for flights); and, of course, the vagaries of mountain weather.

The researchers wanted to cover an area of several square miles, in as much detail as possible. We briefly discussed using a fixed-wing drone because of the large area that needed to be covered with each flight, but this was discounted due to the lack of flat areas for landing and the weight and bulk of the aircraft. Instead, we opted for a small multirotor platform (DJI Phantom 3 Pro), which we knew was reliable and would integrate well with the mapping software (Pix4D).

After some checking with the relevant authorities in Spain we found that we would need permits—not only for the research work, but also for the drone, as flying UAVs is not usually permitted within the Picos area. Furthermore, when we checked with the Spanish aviation authority we discovered that a commercial drone operator needs to have fulfilled all of the local requirements and that my UK qualification and "Permission for Aerial Work" was not recognized in Spain. Fortunately we were able to locate a local pilot who was familiar with the mapping software and had a Phantom 3.

Once in the area of interest, we unpacked the drone and the tablet (which was preloaded with a cached copy of the satellite map of the area). The area to be surveyed was plotted out in the map, ensuring that the flight altitude would give clearance from surrounding mountain peaks, and that the photo overlap was sufficient for our stitching purposes. With relatively low temperatures—particularly early in the morning—we were careful to keep the batteries warm in an insulated backpack and allow the drone sufficient time to warm up.

After a quick weather check for wind and approaching cloud, the pilot made a manual take off, and then initiated the Pix4D automated mission. It's a strange and slightly unnerving experience to watch a valuable drone fly off into the distance with no one at the controls. However, we took comfort from the fact we could watch the drone's progress—both as a speck in the sky and as a dot on the tablet screen—as it worked its way backward and forward, with a reassuring click each time it took a photograph.

Once the mission was complete, the drone obediently returned to its starting point and began a slow descent, at which point the pilot took manual control for a safe landing.

As we were planning to map an area much greater than one flight grid could cover, we had to plan a pattern of grids to cover the area of interest and allow sufficient overlap so that we could stitch all of the resulting grids together. Depending on the size of each grid, sometimes it was possible to fly two grids on one battery, but we were always planning conservatively, so the drone would land with 30% or more still in the battery.

When possible, we also planned our take off and landing spot to be among the higher points in the grid, so we didn't have to worry as much about the drone automatically flying itself into a hillside. Although the drone we were using knew how high above the take off point it was, it didn't have terrain maps built into it, so there was nothing to stop it flying into the steep mountainous terrain. (While newer models do have some terrain obstacle avoidance, I still wouldn't have liked to rely on it, particularly when it's on an automated mission.)

Most of our missions were flown at the maximum permissible altitude of 400ft (120m). We could have flown lower, for a higher

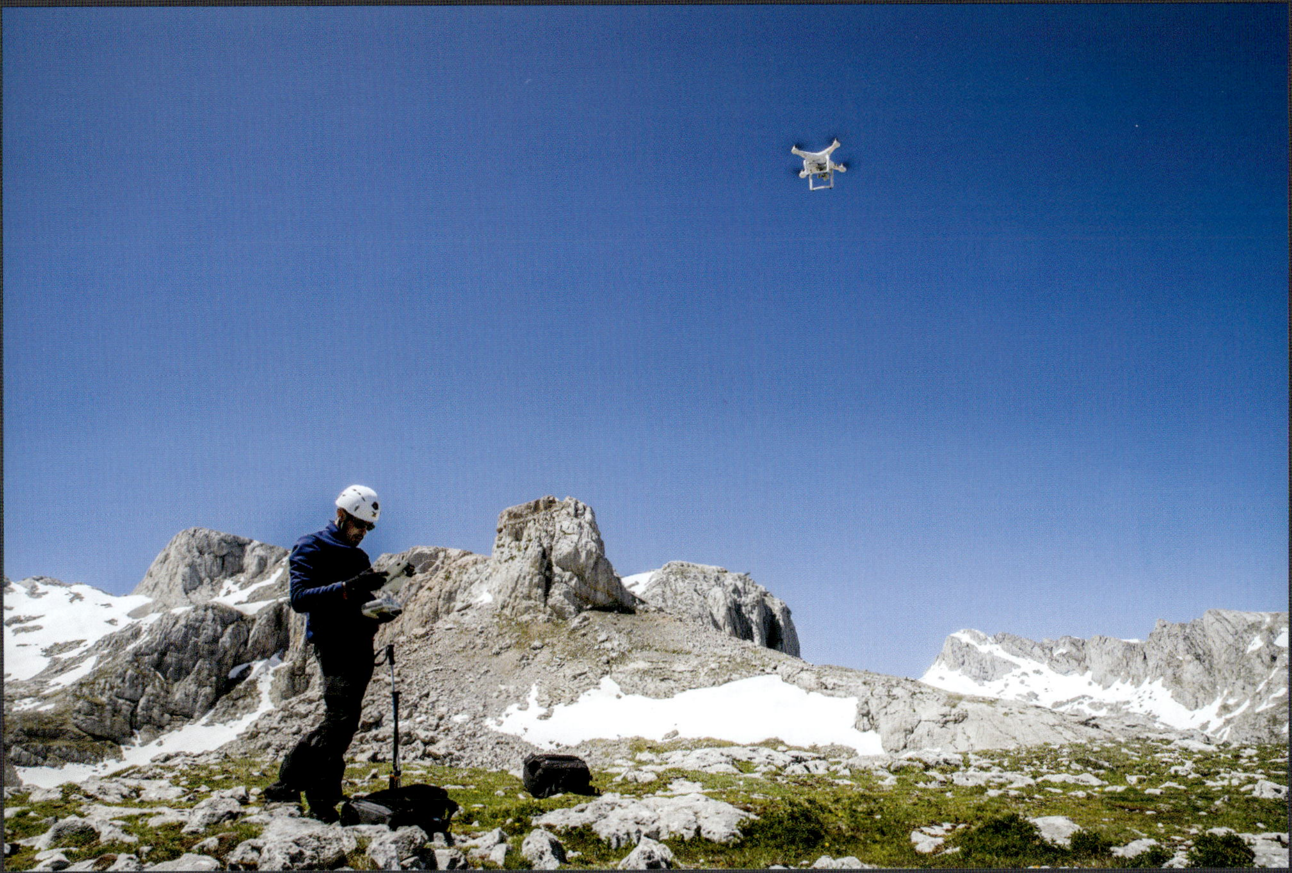

▲ Tony (our drone pilot) in action in the Picos de Europa, as the little Phantom starts another grid mission.

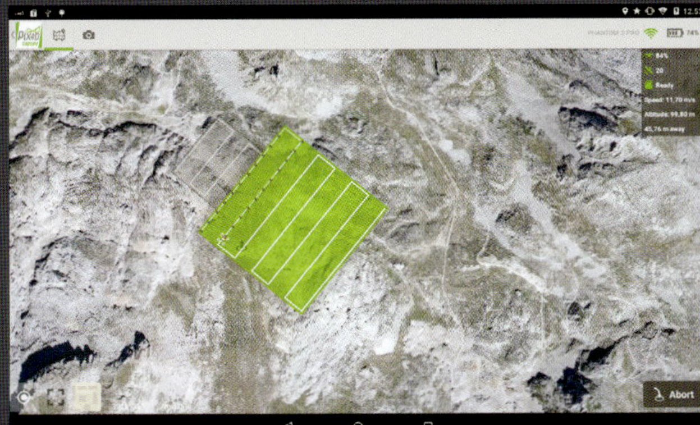

▲ The Pix4D app showing the drone progressing along the grid mission, taking photographs as it goes.

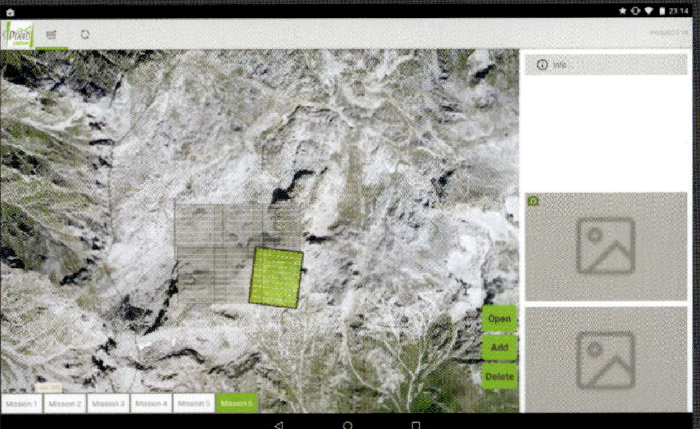

▲ As a large area was being mapped, multiple overlapping grid flights needed to be made.

spatial resolution (giving us more detail in the resulting model), but that would have meant flying a greater number of grids to cover the same area. Even at 400ft there were several occasions when we had to abort a grid flight because it looked as though the drone might collide with the hillside (we were probably safe, but from a distance it is very hard to judge the proximity of the drone to the land around it).

As we were in mountainous terrain, there were several parts of the survey area where we couldn't record sufficient detail with a directly downward-looking viewpoint. As the geologists were interested in the rock formations exposed on the cliff faces, it meant we also needed to fly some manual ("freeflight") missions, with the camera set to fire automatically at appropriate horizontal and vertical distance intervals. Effectively, where we had a cliff, we stood at the top and manually flew a grid on a vertical plane.

As this was a multi-day expedition, we checked each day's results in the evening, using rough, low-resolution processing so we could quickly tell how well the day's flights had worked. To produce a high-resolution model we had to leave the computers running overnight.

However, the main post-processing challenge was the inherent inaccuracy of the altitude readings on the drone. GPS vertical accuracy is inherently worse than its horizontal accuracy, so the drone also makes use of a barometer to detect changes in altitude. As barometric pressure is changing slowly almost all the time, we often found that the terrain on adjacent grids surveyed on different days was vertically displaced in the resulting model. To correct this, we had to specify a set of points ("Ground Control Points," or GCPs) that were visible in the overlap area on both grids and assign latitude, longitude, and elevation to each of them, which we recorded with a handheld GPS unit.

As a result of the Picos drone-mapping project, the geologists ended up with a high-resolution 3D map of a very mountainous area. They were in the field during the drone survey, so they were still able to do a good amount of "ground trothing" by getting down on their hands and knees with their hammers to identify rocks. When this was done in conjunction with a GPS waypoint, it meant they could later correlate a particular rock to a point on the map. Having the geologists in the field also allowed us to make changes to the plan—if they saw a particular area of interesting-looking rock that was not visible on Google Earth or any of the other maps we were using, we could redirect the drone survey.

Dr Veerle Vendeginste, the project's director, summarized the drone mapping like this: "Over a seven day period we hiked around 50 miles (80km) and took 5359 drone photographs during 80 flights. The resulting maps covered an area of just over 2.5 square miles (4km^2) at a maximum spatial resolution of around 4in (10cm). This compares very favorably to the previous year's survey when—working on foot and mapping in the traditional manner—we only managed to map 0.5 square miles (0.75km^2) in a two-week period. The drone also gave us access to more inaccessible areas, such as mountain tops and cliff faces."

TRICKY TERRAIN

During this survey, snow cover on the mountainous terrain created several problems. Not only did it cover up some of the geological features of interest, but the modeling software also needs to identify the same points ("tie points") in several adjacent, overlapping photographs. If the source photographs were predominantly flat snow, the software could struggle to identify sufficient points. In this instance, we were lucky—enough of the snow had melted that only patches remained and each photograph had enough features in it to allow stitching and point cloud generation. Similar problems can be encountered with areas that have large bodies of water within them.

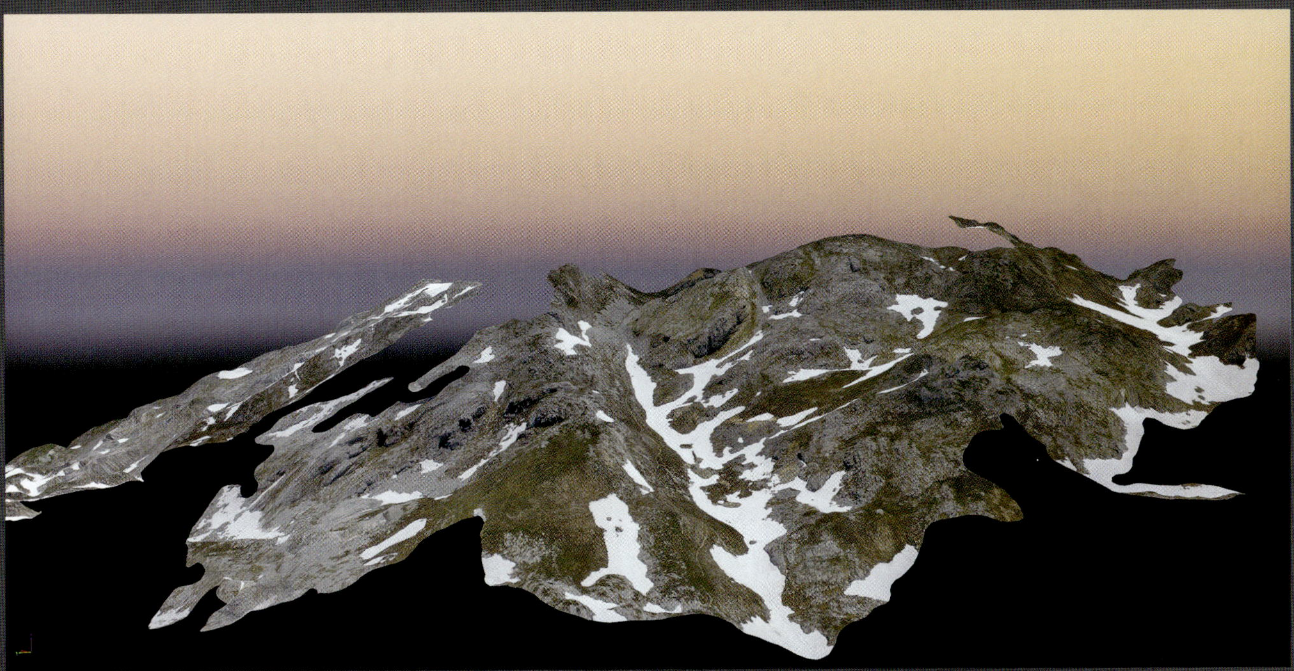

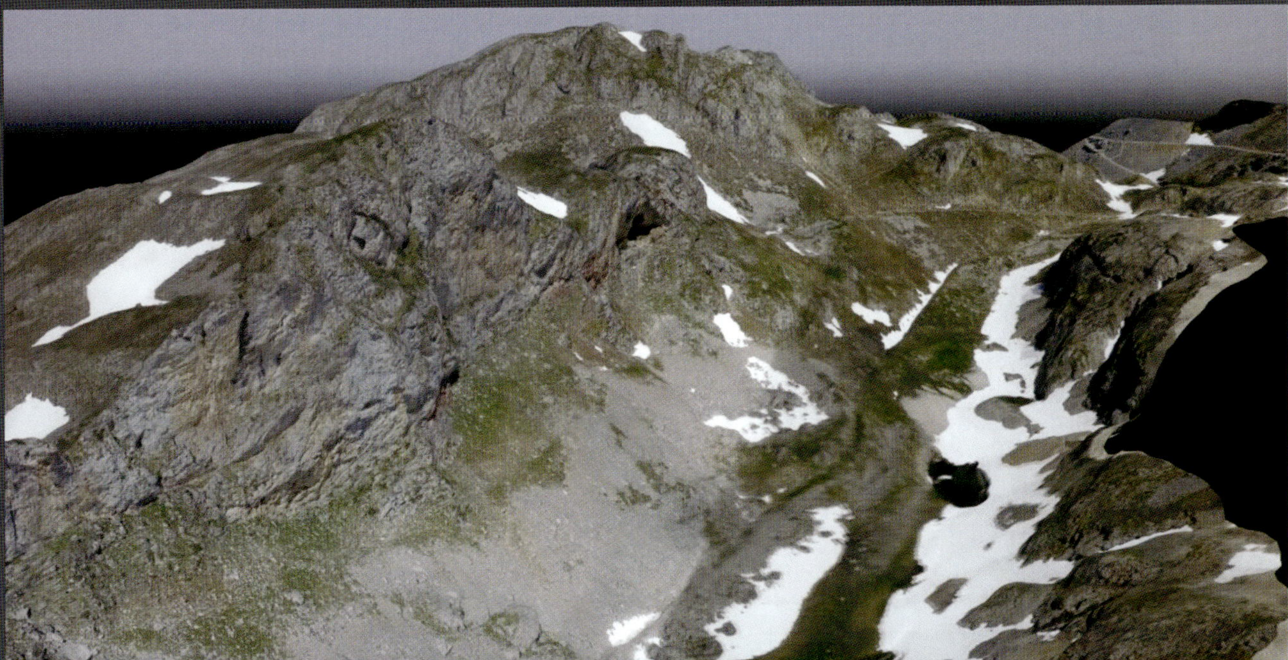

▲ On uneven terrain you need good coverage from a variety of angles to avoid blank areas in your 3D map. Sometimes this can be achieved by running overlapping grids with the camera pointing along both the x axis and y axis of the grid. Other times it may be necessary to take some oblique shots as well, which is what was needed here to fill in the detail on the cliff face at the left side of this 3D model.

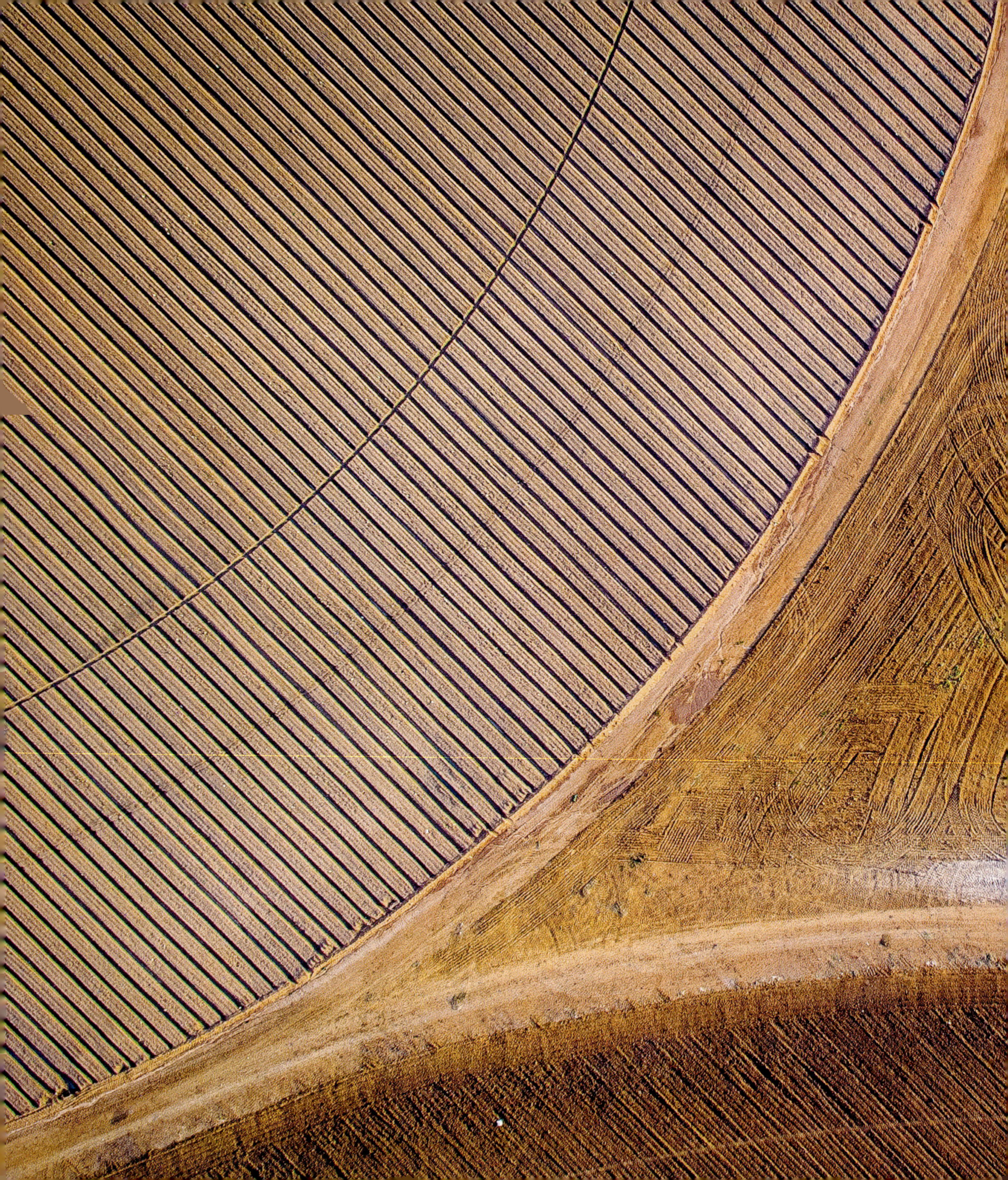

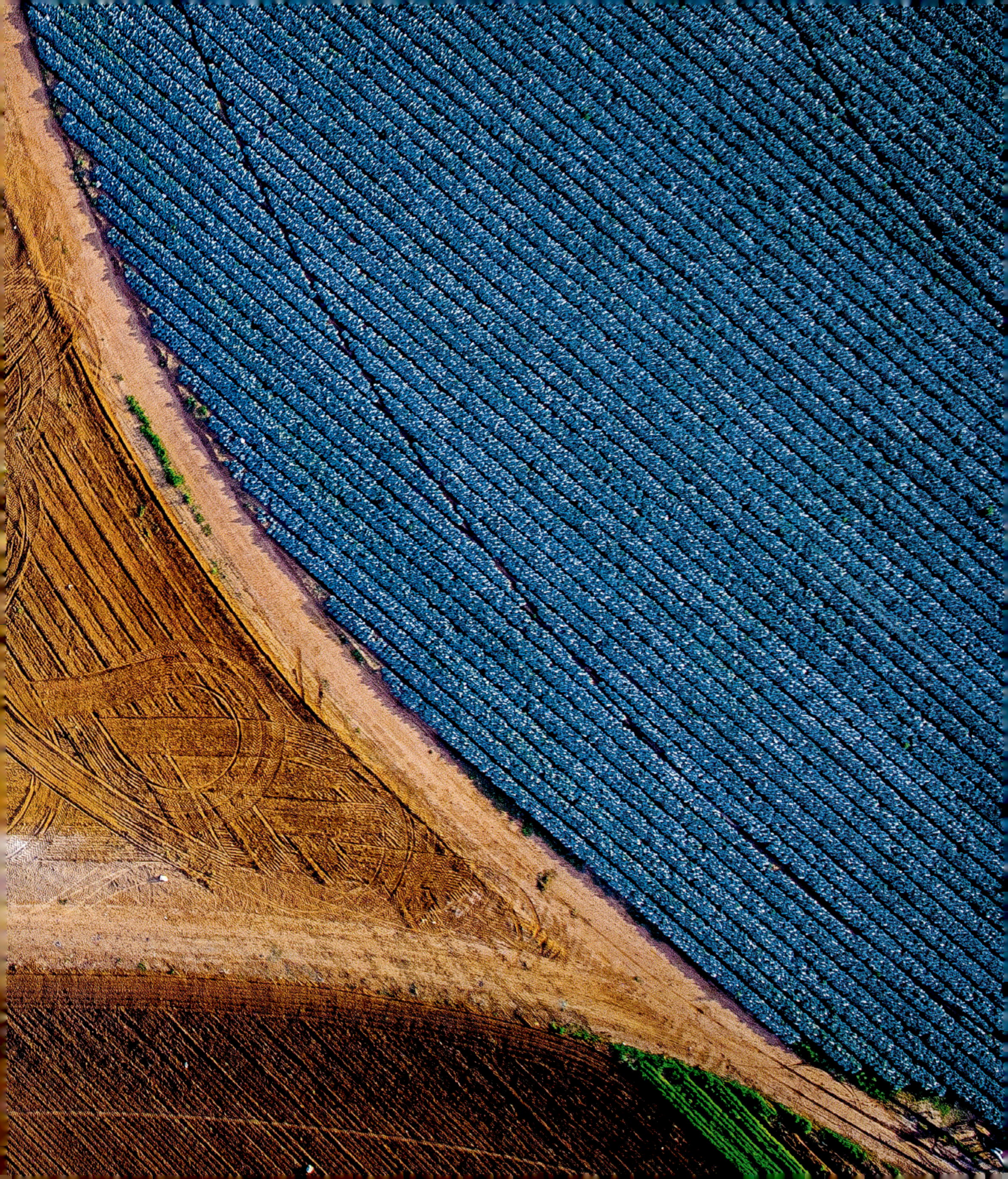

GLOSSARY

Accelerometer: An electronic device for measuring acceleration (usually used on all three planes of movement). An accelerometer is vital in allowing a drone to maintain its orientation and attitude while flying.

AGL (Above Ground Level): The height of the aircraft above the ground directly below it.

Barometric pressure sensor: A component found on many drones that uses air pressure to gauge the aircraft's height above its take off point.

CAA: The United Kingdom's Civil Aviation Authority.

Commercial flight: Any flight made with the purpose of making money.

Drone: An unmanned aerial vehicle (*see also* UAV).

PAGE 168–169: This abstract image was taken by Eddie Oosthuizen as new crops were sprouting after a very cold winter. The crops are planted in circles for efficient irrigation; this triangle between them is only visible from the air.

ESC (Electronic Speed Controller): A component that governs the speed of each flight motor. There is one ESC per motor.

FAA: The United States Federal Aviation Administration.

Firmware: The software that comes preloaded on a drone or remote control. This may need to be updated periodically to fix bugs or add new features.

FPV (First Person View): a way of piloting a drone whereby the pilot relies primarily on the view through the drone's onboard camera, often viewed through special goggles.

Flight controller: The "brain" of a drone. Often abbreviated to FC, the flight controller keeps the drone stable and orchestrates its movements, with input from the pilot, the accelerometers, compass and GPS units, and the app.

Flyaway: A situation where the pilot unintentionally loses control of the drone, leading to it flying away.

Geofence: A virtual fence that can be used to prevent a drone from taking off inside—or entering—restricted areas. They rely on GPS coordinates.

Gimbal: A stabilized camera support, usually run from battery power, which uses motors and accelerometers to keep the camera stable and respond to commands from the operator(s).

GLONASS: The Russian equivalent to GPS (*see below*). Many drones can use both GLONASS and GPS systems.

GPS (Global Positioning System): A system of orbiting satellites that allow earth-based devices to calculate their position to within a few feet/meters. Also allows a drone to maintain its position, despite wind. However, as the drone must have substantial sky visibility to get a position fix, GPS doesn't work well indoors.

Gyroscope: Electronic component that—along with accelerometers—provides information to the drone on its force and acceleration status.

Hexacopter: Multirotor aircraft with six rotors.

Home point: The point to which the drone will return in the event of signal loss or low battery. This may be the take-off point, or it may be dynamic (moving with the pilot).

IMU (Inertial Measurement Unit): Used to measure a drone's specific force and angular rate, using a combination of accelerometers and gyroscopes.

IOC (Intelligent Orientation Control): DJI's terminology for a series of different flight modes. In these modes, the heading of the front of the aircraft does not necessarily correspond to the forward direction of the right control stick.

JPEG: A compressed image file format.

LIPO (Lithium Polymer): a battery technology commonly used in drones and other electronic devices.

LOS (Line of Sight): Keeping the drone within the viewing range of the naked eye.

Multirotor: An aircraft with more than two main rotors.

NAA (National Aviation Authority): The organization responsible for air safety in a particular country (for example, the FAA in the United States or the CAA in the United Kingdom).

Octocopter: A multirotor aircraft with eight rotors.

OSD (On Screen Display): Telemetry from the drone displayed on a monitor. Data might include the drone's height, battery level, and distance from user, among other things.

Payload: The combined weight of anything the drone is carrying, such as the camera, lens, and gimbal in the case of a camera drone.

Phantom: A popular line of camera drones from DJI.

Pitch: A movement of the drone along an axis whereby the nose of the aircraft points down or up relative to horizontal.

POI (Point Of Interest): A specified geographic point, about which the drone makes an orbital flight, with the camera oriented so as to keep the POI central in the frame.

Quadcopter: A multirotor aircraft with four rotors.

Raw: An uncompressed file format for photographs or video that offers room for adjustment in post-production. Raw files take up more memory card space than compressed files such as JPEG.

RC (Radio Control): Often used to refer to the radio control transmitter unit.

Roll: Rotation of the aircraft around its nose-to-tail axis

RPAS (Remotely Piloted Aircraft Systems): Term occassionally used to refer to a drone and its associated control and monitoring system.

RTH (Return To Home): Feature whereby the drone automatically returns to a specified point (often its take off point or pilot location) in the event of signal loss or low battery.

Rx: Receiver.

SUAS (Small Unmanned Aircraft Systems): Term occassionally used to refer to a drone and its associated control and monitoring system.

Telemetry: Flight data transmitted by the aircraft.

Tx: Transmitter.

UAV (Unmanned Aerial Vehicle): Term commonly used to refer to drones.

Waypoint: A set of GPS coordinates used in autonomous navigation.

Yaw: To rotate the aircraft on its own axis on a horizontal plane.

USEFUL WEBSITES

Safety & legal

National Aviation Authorities: www.en.wikipedia.org/wiki/List_of_civil_aviation_authorities

FAA (United States): www.faa.gov/uas/

CAA (United Kingdom): www.caa.co.uk/Consumers/Model-aircraft-and-drones/Flying-drones/

DGAC (France): www.developpement-durable.gouv.fr/politiques/aviation-civile

LBA (Germany): www.lba.de/EN/Home/home_node.html

AESA (Spain): www.seguridadaerea.gob.es/LANG_EN/home.aspx

Drone laws (country by country): www.uavcoach.com/drone-laws/

Hardware

3D Robotics: www.3dr.com/solo-drone/

AirSelfie: www.airselfiecamera.com

Align: www.align.com.tw/multicopter-en/

Blade: www.bladehelis.com/Aircraft/

Cheerson: www.cheerson.com

DJI: www.dji.com

Freefly: www.freeflysystems.com

Hubsan: www.hubsan.com

Intuitive Aerial: www.intuitiveaerial.com

JJRC: www.jjrctoy.com

Parrot: www.parrot.com

Powervision: www.powervision.me

Sensefly: www.sensefly.com/home.html

Tarot: www.tarot-rc.com

Yuneec: www.yuneec.com/en_GB/home.html

Zerotech: www.zerotech.com

Software

Adobe (Lightroom, Photoshop & Premiere Pro): www.adobe.com

Agisoft: www.agisoft.com

Altizure: www.altizure.com

Apple Final Cut Pro X: www.apple.com/uk/final-cut-pro/

DJI (software & firmware updates): www.dji.com/support/product

Drone Deploy: www.dronedeploy.com

Pix4D: www.pix4d.com

Windguru: www.windguru.cz

Apps

AirMap Airspace Information (primarily US): www.airmap.com

Autopilot: www.autoflight.hangar.com

DJI Go: www.dji.com/goapp

Drone Assist (primarily UK): www.dronesafe.uk/drone-assist/

Hover (weather & airspace information): www.hoverapp.io

Litchi: www.flylitchi.com

The Photographer's Ephemeris: www.photoephemeris.com

Support & forums

DJI Forums: www.forum.dji.com

Inspire Pilots: www.inspirepilots.com

Mavic Pilots: www.mavicpilots.com

Phantom Pilots: www.phantompilots.com

RCGroups: www.rcgroups.com/forums/index.php

DJI YouTube tutorials: www.youtube.com/channel/UClH0xVO3zOfYdGjoPU6S2hw

Inspiration & sharing

AirVuz: www.airvuz.com

Dronestagram: www.dronestagr.am

Epic Drone Videos (YouTube channel): www.youtube.com/channel/UC9FmF7MZlsl3QCWtuCAnOeQ

London Drone Film Festival: www.londondronefilmfestival.com

New York City Drone Film Festival: www.nycdronefilmfestival.com

Skypixel: www.skypixel.com

Official DJI Owners Group: www.facebook.com/groups/dji.owners/

Photographers

Felix von Aulock: www.flickr.com/photos/106011035@N05/

Alex Baker: www.droid-cam.com

Fredrik Christiansen: www.mucru.org/whale-study-using-drones-and-tags-captures-aerial-footage-of-white-whale-calf/

Andy Deitsch: www.flickr.com/photos/andydeitsch/albums

Fly Through Films: www.flythroughfilms.com

David Hopley: www.drawswithlight.co.uk

Fergus Kennedy: www.ferguskennedy.com

James Loveridge: www.jamesloveridgephotography.co.uk

Eddie Oosthuizen: www.skypixel.com/user/eduardt

Lec Park: www.lecpark.com

Skylark Aerial Imaging: www.skylarkaerialimaging.com

Miguel Willis: www.miguelwillis.com

Andy Yeung: www.andyyeungphotography.com

PICTURE CREDITS

All photographs © Fergus Kennedy, with the exception of the following:

Aerial Motion Pictures 50; AirSelfie 26; Felix von Aulock 94–95; Alex Baker 147; Brooklyn Museum, Frank L. Babbott Fund 12; Ajit Chambers 10; Fredrik Christiansen 84–85; Andy Deitsch 77 , 113–115; Fly Through Films 33T, 38, 53; David Hopley 130–131; Nick Key 30; Library of Congress 13, 17; James Loveridge 149; Eddie Oosthuizen 44–45, 168–169; Lec Park 64BL, 154–155; Nick Park 65BL; Powervision 32; U.S. Air Force Photo/Lt. Col. Leslie Pratt 15; Sensefly 156–157, 159, 161; Arun Taylor 72; Miguel Willis 65TR, 70–71; Alex Wykes 82, 92; Andy Yeung: 2, 20–21

INDEX

ABOUT THE AUTHOR

An early passion for the outdoors led Fergus Kennedy to train as a marine biologist: after studying Zoology at Oxford University he gained a PhD in Coastal Ecology from the University of North Wales, whilst doing fieldwork in Chile. Photography was a key part of Fergus' work, and this developed in a number of directions—from land-based imaging to underwater photography.

When drones appeared on the scene, Fergus immediately saw their potential. Having gained his CAA permission, Fergus set up Skylark Aerial Imaging, which opened the door to professional aerial work. He has shot aerial video for many television channels, including ABC, the Discovery Channel, the BBC, ITV Drama, Channel 4, and Channel 5. Fergus has also undertaken drone-based aerial still photography for real estate, architecture, and geological surveys, and produced numerous commercial projects for clients such as Canon Europe, WWF, and Toyota. In this time he has flown everything from tiny microdrones to heavy lift, 12 rotor drones carrying cinema cameras.

However, the challenge of dreaming up and executing personal photographic projects still gives Fergus the greatest pleasure. His work has been Highly Commended in the *Wildlife Photographer of the Year* competition and his images have been published in a numerous books, as well as featuring in National Geographic's *Wild Arabia* exhibition. Fergus is also a frequent contributor to *Outdoor Photography Magazine* and sits on the judging panel of the *Outdoor Photographer of the Year* competition.

ACKNOWLEDGMENTS

Firstly I'd like to thank those friendly and helpful folks from Ammonite Press—Jason Hook and Robin Shields—who encouraged and steered me through the turbulence and into clear skies, Chris Gatcum for editing it all, and Luke Herriot for doing a wonderful job on the layout.

I wouldn't have become the drone nerd I am today without help from those who had gone before: Alex Baker of Droid-Cam gave me lots of encouragement in the early days, and Leo Bund and Henry Bridges of Fly Through Films provided a fantastic insight into the world of high end aerial film work. I also want to thank Saffron Allwood, Alex Wykes, Nick Kennedy of Verri Media, Hugh Fox, and the inimitable Bob Jaroc, who were always on hand with helpful comments (some of them even constructive!), as well as ensuring a steady flow of coffee and cookies.

Finally, my family—always there, and rarely getting the thanks they deserve. To my parents who gave me a wonderful upbringing and continue to support me: my dad, Finn, for helping out with crazy drone projects and customizing parts in the workshop, and my mum, Anne, for looking after our kids and cooking fantastic Sunday lunches. My wife, Anna, who has tolerated my droning on (sorry!) with minimal eye-rolling, as well as being an expert editor, and our two sons—Alfie and Harry—who brim with enthusiasm, even after a lengthy trek to a muddy and windswept beauty spot!

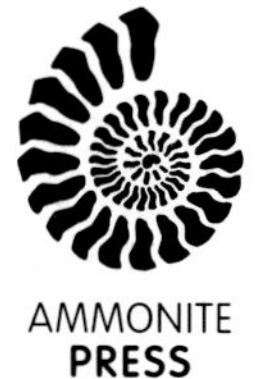

AMMONITE
PRESS

www.ammonitepress.com